Bioprocesses and Engineering

Bioprocesses and Engineering

Managing Editor: A. Fiechter

with 55 Figures and 24 Tables

Akademie-Verlag Berlin

Die Originalausgabe erscheint im Springer-Verlag Berlin · Heidelberg · New York
als Volume 40 der Schriftenreihe
Advances in Biochemical Engineering/Biotechnology

Vertrieb ausschließlich für die DDR und die sozialistischen Länder: Akademie-Verlag Berlin

ISBN 3-540-50151446-5 Springer-Verlag Berlin Heidelberg New York
ISBN 0-387-50151446-5 Springer-Verlag New York Berlin Heidelberg

ISBN 3-05-500803-0 Akademie-Verlag Berlin

Erschienen im Akademie-Verlag Berlin, Leipziger Straße 3—4, Berlin, DDR-1086
Lizenznummer: 202 · 100/555/89
Printed in the German Democratic Republic
Gesamtherstellung: VEB Druckerei „Thomas Müntzer“, 5820 Bad Langensalza
LSV 1315
Bestellnummer: 764 111 2 (3070/40)
12800

Table of Contents

Recent Developments in Enzyme and Microbial Biotechnology — Strategies in Bioprocess Design
N. Hardman 1

Protein Purification
E. P. Lillehoj, V. S. Malik 19

Lipid Biosynthesis in Oleaginous Yeasts
G. Turcotte, N. Kosaric 73

The Use of Calorimetry in Biotechnology
U. von Stockar, I. W. Marison 93

Effect of Total and Partial Pressure (Oxygen and Carbon Dioxide) on Aerobic Microbial Processes
U. Onken, E. Liefke 137

Recent Developments in Enzyme and Microbial Biotechnology — Strategies in Bioprocess Design

Norman Hardman
Department of Molecular Biology, Biotechnology Section, K-681.4.43,
CIBA-GEIGY AG, CH-4002 Basel, Switzerland

1 Introduction ... 1
1.1 Historical Perspective ... 1
1.2 Growth of Microorganisms in Bioreactors ... 1
1.3 Bioprocess Control ... 1
1.4 Impact of Molecular Biology ... 3
2 Chemical Synthesis Using Enzymes from Microbial Sources ... 3
2.1 Specific Chemical Transformations ... 3
2.2 Current and Future Developments ... 4
3 Recent Advances in Protein Structure Determination and its Biotechnological Applications ... 5
3.1 X-ray Diffraction ... 5
3.2 Nuclear Magnetic Resonance ... 6
3.3 Structure of Rare Proteins ... 6
3.4 Mutational Analysis of Protein Structure: Protein Engineering ... 7
3.4.1 Site-directed Mutagenesis ... 7
3.4.2 Predicting Changes in Protein Structure ... 7
3.4.3 Antibody Engineering ... 8
3.4.4 Quality Control in Production of Recombinant Proteins ... 8
4 Sensors, Biosensors and Their Applications ... 10
4.1 Concept and Applications ... 10
4.2 Structure of Biosensors ... 11
4.2.1 General Aspects ... 11
4.2.2 Biological Components ... 12
4.2.3 Transducers ... 13
4.3 DNA Sensors ... 14
5 Addendum ... 16
6 References ... 17

Microorganisms have been used traditionally by industry as sources of natural products, or as sources of enzymes capable of mediating specific chemical transformations. This situation has changed radically in recent years, a time during which we have seen a dramatic increase in the number and range of potential biotechnological applications of enzymes and their genetically-engineered variants. An increasing number of enzymes, receptors and other proteins have now been structurally characterized, and their genes isolated as a basis for producing recombinant proteins for genetic analysis of their structure and function. These innovations have necessitated development of associated technologies for large-scale production of proteins in bioreactors, appropriate strategies for quality control, and new analytical tools for structural characterization of recombinant gene products. Some recombinant proteins are already in an advanced stage of development for use either as new-generation therapeutics, as target molecules for "intelligent" drug screening, or as biological components of biosensors. As the predictive power of protein model building improves, the diversity of applications of such technology will increase further as it becomes feasible to generate totally synthetic proteins with specifically-tailored properties.

1 Introduction

1.1 Historical Perspective

"Biotechnology", in the broadest sense of the word, is almost as old as civilization itself. For thousands of years man has made use of naturally-occurring microorganisms to produce foodstuffs and beverages. The plant kingdom has additionally provided a wealth of pharmacologically active substances of medicinal value, almost all of which were used to advantage at the time without knowledge of their mechanism of action. These early "biotechnological" activities are a tribute to man's ability to control his environment through ingenuity and the adventitious use of available natural resources.

1.2 Growth of Microorganisms in Bioreactors

In modern times we have seen a considerable expansion in the scale, scope and innovative uses of biotechnology, paralleled by a more rational approach to biotechnologically-important questions as it became possible to grow the relevant microorganisms under controlled conditions on an industrial scale. Microbiology provided the means to identify and cultivate a wide range of hitherto uncharacterized organisms capable of producing novel substances with potent biological activities, such as antibiotics. Organisms have been discovered with enzymes which can generate efficiently important substances such as vitamins and amino acids from simple substrates, and yet other organisms have been found which can bring about relatively complex, stereospecific chemical transformations with astounding efficiency. Biochemistry has been important in defining the nature and complexity of the biosynthetic pathways which lead to a particular natural product, and studies of the physiology of cell growth have been essential for optimizing production rates of these substances in bioreactors. Finally, classical genetics has proved to be an invaluable means of generating new mutant cell lines optimized for product yield and large-scale cultivation.

1.3 Bioprocess Control

These essential contributions of the Biological Sciences to the development of modern Biotechnology have been matched by an equally significant contribution from the Physical and Engineering Sciences. New engineering problems have been addressed in the design of efficient production plants for large-scale cultivation of microorganisms. It has become necessary to develop novel, sensitive sensor devices for monitoring rapid changes in biological parameters such as growth rate, pH, oxygen tension and nutrient supply in order to optimize microbial growth for maximal production of a substance. The use of bioreactors to generate chemical substances and more complex biomolecules has also presented entirely new problems associated with product purification, treatment of biomass and waste disposal.

1.4 Impact of Molecular Biology

The last decade has seen a further biotechnological revolution following the application of knowledge gained from Molecular Biology. When examined in a historical context, this can be seen as an additional extension of the contribution that advances in basic knowledge in the Biological Sciences have made to biotechnological innovation. Nonetheless, these recent advances have added a new dimension to the application of biotechnology within the pharmaceutical and chemical/agrochemical industries. For the first time we are presented with the possibility of creating new, specifically engineered macromolecules with desirable biological properties either for use as therapeutics, as tools for the development of novel bioprocesses and construction of biosensors, or as a basis for development of novel and efficient screening systems for drug discovery.

The following review summarises some advances in several of the above areas of biotechnology and bioprocess development, based partly on the proceedings of a recent Meeting [1].

2 Chemical Synthesis Using Enzymes from Microbial Sources

2.1 Specific Chemical Transformations

Many microorganisms contain specific enzymes capable of directing particular chemical transformations, the most well-known being those leading to the synthesis of amino acids and cofactors. As in the case of antibiotic-producers, natural habitats have traditionally provided a rich source of the relevant microorganisms. In some parts of the world this concept is no longer considered to be as fashionable as previously, but in Japan this painstaking but pragmatic, intuitive approach has continued to lead to the identification of a range of new and useful enzymes from a diverse selection of organisms of different genera (Table 1), and the list continues to grow yearly. In some instances different organisms can be found containing enzymes capable of efficiently synthesising alternative stereochemical forms of a compound (e.g. synthesis of D- or L-cysteine; Table 1). In other cases growth conditions for the same organism can be manipulated to produce different enzymes. *Rhodococcus rhodochrous*, for example, produces either nitrile hydratase or nitrilase, depending on the growth conditions employed; the addition of crotonamide and cobalt chloride selectively induces the production of nitrile hydratase, whereas isovaleronitrile or isobutyronitrile are potent inducers of nitrilase. These, and other organisms such as *P. chlororaphis* and *Brevibacterium* can be used to produce a range of amines with diverse applications, such as acrylamide or nicotinamide and biologically-active amines such as L-Dopa [2–5] (Table 1). Many of these compounds are important products in their own right, whereas others are important stereospecific intermediates required for further steps in chemical syntheses of additional compounds.

Table 1. Chemical production using enzyme synthesis. The Table shows a representative collection of chemical compounds generated by enzyme synthesis, together with yields. The functional enzyme and bacterial source are indicated. (Table courtesy of Dr. H. Yamada, Laboratory of Agricultural Chemistry, Kyoto University, Kyoto, Japan)

Product	Enzyme (source)	Yield	
		$g\,l^{-1}$	mol%
Amino acids			
D-*p*-Hydroxyphenylglycine	Dihydropyrimidinase (*Bacillus* sp.)	4.9	(74)
D-Phenylglycine	Dihydropyrimidinase (*Bacillus* sp.)	6.2	(91)
L-Tyrosine	β-Tyrosinase (*Erwinia herbicola*)	61	
L-Dopa	β-Tyrosinase (*Erwinia herbicola*)	53	
L-Tryptophan	Tryptophanase (*Proteus rettgeri*)	100	(95)
L-Cysteine	Cysteine desulfhydrase (*E. cloacae*)	50	(86)
L-Cysteine	Cysteine synthase (*B. sphaericus*)	70	(82)
D-Cysteine	β-Chloro-D-alanine lyase (*P. putida*)	22	(88)
L-Cystathionine	Cystathionine γ-synthase (*B. sphaericus*)	42	(92)
L-Serine	Serine transhydroxymethylase (*Hyphomicrobium* sp.)	35	(25)
R-4-Chloro-3-hydroxy-butyric acid	Aldehyde reductase (*Sporoboromyces salmonicolor*)	72	(95)
Amides			
Acrylamide	Nitrile hydratase (*P. chlororaphis*)	400	(98)
Methacrylamide	Nitrile hydratase (*P. chlororaphis*)	200	
Crotonamide	Nitrile hydratase (*P. chlororaphis*)	200	
Nicotinamide	Nitrile hydratase (*Rhodococcus rhodochrous*)	1465	(100)
Nicotinic acid	Nitrilase (*Rhodococcus rhodochrous*)	172	(100)
Pyrogallol	Gallic acid decarboxylase (*Citrobacter* sp.	23	(100)
Theobromine	Oxygenase (*P. cepacia*)	14	(72)
D-Pantoyl lactone	Carbonyl reductase (*Candida parapsilosis*)	100	(83)
Coenzymes			
Coenzyme A	Multi-step enzyme system (*Br. ammoniagenes*)	115	(100)
S-Adenosylmethionine	AdoMet synthetase (*Saccharomyces sake*)	12	(45)
S-Adenosylhomocysteine	AdoHcy hydrolase (*Alcaligenes faecalis*)	74.2	(97)
FAD	FAD pyrophosphorylase (*Arthrobacter globiformis*)	18	(28)
Pyridoxal 5′-phosphate	PMP oxidase (*P. fluorescens*)	0.15	(98)
NADH	Formate dehydrogenase (*Arthrobacter* sp.)	30	(90)
NADPH	G6P dehydrogenase (a methanol-utilizing bacterium)	7	(75)
Polyunsaturated fatty acids			
Dihomo-γ-linolenic acid	Multi-step conversion (*Mortierella alpina*)	2.2	
Arachidonic acid	Multi-step conversion (*Mortierella alpina*)	3.6	
Eicosapentaenoic acid	Multi-step conversion (*Mortierella alpina*)	1.4	

2.2 Current and Future Developments

Current emphasis is being placed on development of flexible, efficient multi-step processes involving both enzymes and conventional chemistry. This is a good example of the synergy between biology and traditional industrial chemistry which can lead to novel solutions to problems associated with particular complex organic syntheses. Japan currently has one of the leading positions in this new biotechnological approach.

A number of additional future developments could be envisaged in this area of Biotechnology. Apart from strain improvement using classical genetics or recombinant genetics, where such industrially-important enzymes are in short supply the techniques of Molecular Biology could be used to clone the relevant gene and over-produce the enzyme in appropriate homologous or heterologous genetic expression systems. Knowledge of the structure of such proteins could also be used to generate genetically-engineered variants with improved properties specifically tailored for application in bioreactors. These might include molecules with improved heat-stability or altered selectivity within a given range of substrates. Structural studies have already been initiated on a number of biotechnologically-important enzymes identified as potential targets for this type of protein engineering [6-10].

3 Recent Advances in Protein Structure Determination and its Biotechnological Applications

3.1 X-ray Diffraction

Protein structure determination has for many years provided an important tool to aid drug discovery in the pharmaceutical industry. Where an enzyme has been identified as a potentially important biochemical target, knowledge of the stereochemical configuration of its active-site and the location of important amino acid residues involved in the enzymic mechanism can provide the critical information necessary for the design of highly-specific chemical inhibitors.

X-ray diffraction of single crystals of a purified protein has been the traditional method of obtaining this three-dimensional structural information, once the primary amino acid sequence was known. During the early phase of development of this technology in the late 1950s and 1960s, X-ray structures of proteins accumulated slowly, since available methodologies were rather slow and unsophisticated by present-day standards. Primary sequence information was obtained by painstaking analysis of overlapping polypeptides purified after proteolytic or chemical cleavage; reliable methods were lacking for the generation of protein crystals appropriate for X-ray diffraction analysis; data collection techniques and computing facilities were primitive. Dramatic improvements have been made in virtually all these aspects of X-ray diffraction technology, so that today X-ray analysis of much larger, complex protein structures and protein assemblies can be attempted with a high degree of success. The time-scale required to complete such structural analyses has also been reduced considerably as a consequence; in contrast to early studies taking several years to complete the most recent innovations permit the elucidation of the structure of an "average" protein in favourable cases in less than six months. Additionally, the advent of recombinant DNA technology allows the necessary primary polypeptide sequence information to be deduced from the cloned complementary DNA (cDNA) sequence encoding the protein of interest in a fraction of the time needed to obtain similar information directly from the protein itself. The most uncertain step in X-ray diffraction analysis remains that of obtaining suitable single crystals of the purified protein; on occasion crystals can be obtained easily, in other cases it proves difficult or impossible. No hard-and-fast rules appear to govern the suitability of a protein for crystallization; the best crystals often result from a process of trial and error.

3.2 Nuclear Magnetic Resonance

An alternative method of protein structure determination, nuclear magnetic resonance (NMR), is now receiving increased attention. The method yields valuable complementary data to that obtained using X-ray diffraction analysis, providing a greater insight into protein structure/function relationships than previously available using X-ray analysis alone. One major advantage of the approach is that protein specimens are analysed in solution or in other non-crystalline states, avoiding the requirement for single crystals on which X-ray analysis is dependent. For some proteins extensive similarities between the X-ray crystal structure and solution NMR structure are observed [11-13], but in other cases NMR analysis predicts major conformational differences between crystal and solution structures [14-15]. In some instances detailed comparison between crystal and solution structures reveals only subtle conformational differences which, nevertheless, can be significant with regard to the functional properties of the protein [11].

For those proteins where no single crystals are available NMR will, at least in some instances, be the only alternative method to obtain 3-dimensional structural information. The method also has a marked advantage over X-ray crystallography in that NMR data analysis can be performed over a much shorter timescale, and it can provide unique information on the dynamics of protein structure. However, technical constraints currently restrict the application of the NMR method to protein structures with a molecular weight less than 20,000–30,000 Da which can form stable, relatively concentrated solutions (>1 mM [16]).

3.3 Structure of Rare Proteins

Taken together, these recent developments allow X-ray diffraction analysis and NMR to be used as ever more powerful analytical tools for protein structure determination. These methods have already been used to solve the structures of a significant number of proteins of basic biological interest and, increasingly, proteins which are prime targets of pharmaceutical and biotechnological interest as discussed above. However, until recently only naturally-occurring proteins that are relatively easy to isolate and available in sufficient quantity have been amenable to such detailed structural analysis. A small number of additional studies had already been made of rare proteins, obtained only after painstaking purification from large quantities of source material.

Molecular Biology has changed this picture in a revolutionary way. Where a protein/enzyme is available only in minute amounts it is now possible in many instances to clone its gene or cDNA and produce a recombinant form of the protein in non-limiting amounts using a number of alternative genetic expression systems. Thus, for the first time it has become possible to derive information about the structure of a whole range of rare and often complex proteins hitherto unobtainable by conventional means. Examples include cell receptors for biologically and pharmaceutically-important ligands [17,18], target enzymes for the design of specific therapeutics [19], hormones [20], recombinant growth factors [21] and many others.

Efforts worldwide have increased considerably the range of vectors that can be used for the manipulation of recombinant DNA molecules, and the range of organisms that can be used for DNA cloning and expression. These include Gram-negative microorganisms (especially *Escherichia coli*), *Bacillus*, *Staphylococcus*, *Streptococcus*, *Haemophilus*, *Neisseria*, *Cyanobacteria*, and a wide range of other organisms including fungi, yeasts, plants and various animal-cell systems [22]. Additional DNA cloning vectors have also been developed, termed "shuttle vectors", which permit their replication in two or more organisms of different genera [22]. Such vectors have proved useful for several applications, especially for genetic manipulation in *E. coli* of genes from microorganisms that are difficult to handle in the laboratory. This wide choice of alternative expression systems has proved necessary because genetic factors such as codon usage, for example, can effect the efficiency with which a given cloned gene can be expressed in a heterologous expression system. Another way around such problems is to generate a totally synthetic gene with appropriate codon bias to facilitate expression but which nevertheless encodes a polypeptide sequence of identical primary structure to the original protein [9].

3.4 Mutational Analysis of Protein Structure: Protein Engineering

3.4.1 Site-directed Mutagenesis

Recombinant DNA methods have likewise revolutionized mutational analysis of proteins. There is no longer a need to rely on classical genetics to generate informative functional mutants. A complete range of new, fast and often sophisticated experimental techniques permit the generation, at will, of precise site-directed genetic changes in polypeptide sequences such as point mutations and deletions. Together with appropriate recombinant gene expression systems for generating the mutant proteins, and rapid methods for their physical characterization, these new procedures offer incisive new tools for the study of protein structure and function using mutational analysis.

3.4.2 Predicting Changes in Protein Structure

In some cases recombinant genetics leads to mutants which may have a particularly useful property, such as an enzyme with enhanced stability or altered substrate specificity. Synthetic variants of the polypeptide protease inhibitor Eglin-C, for example, have an altered preference for particular proteases [23]. Knowledge of the three-dimensional shape of such proteins is an essential aid to making useful mutants, but even with this information unpredictable changes in structure are sometimes seen [24]. This unpredictability stems at least in part from deficiencies in the current methods for computer modelling, which fail to foresee conformational changes which can occur in the three-dimensional structure of a protein even as a result of the introduction of a single point mutation. However, rapid advances are now being made in the development of new and more powerful computer software systems which can take advantage of the availability of a progressively increasing data base of protein structures for the formulation of reliable empirical rules for model building [25, 26]. The significant financial investment and high level of technical expertise required to ensure the success of this approach has resulted in the development of a number of "Centres of Excellence" for protein engineering, notably in the United Kingdom and in Japan [27], where costs are shared between Government and Industry.

Computer-assisted molecular modelling, although still very much in its infancy, will probably play an increasingly important role in the future in allowing the protein chemist to make more predictable changes in the structure of a given protein/enzyme which can then be exploited for biotechnological purposes. Certain enzymes with improved heat-stability, for example, may be more useful than their natural counterparts for incorporation into biosensor devices (Sect. 4.2.2). It can likewise be envisaged that alterations in other properties such as cofactor requirements, pH optimum and substrate specificity could also have important consequences for bioprocess development.

3.4.3 Antibody Engineering

Antibody engineering is one area where the perceived usefulness of protein structure manipulation has already been realized. It has been possible to exploit the vast amount of basic information available on antibody structure and function to generate many different kinds of engineered molecules. These include chimaeric antibody molecules, in which the antigen-combining domains of the heavy- and light-chain immunoglobulin (Ig) polypeptides from one animal species are combined with Ig structural domains of another [28–30]. It has also proved possible to generate Ig:non-Ig chimaeric protein structures [31, 32] and to graft the relevant oligopeptide segments (complementarity determining segments) responsible for antigen specificity from one antibody on to a second Ig molecule [33]. Additionally, antibodies which recognise transition-state intermediates of simple chemical reactions can be generated and shown to act as primitive enzymes [34, 35]. Manipulation of antibody structures illustrates the potential versatility of protein engineering in general and paves the way for future studies in which it may be possible to create a new generation of entirely synthetic proteins with defined, exploitable properties. Studies on such artificial proteins have already begun [25].

As an illustration of this philosophy, Rees and de la Paz [36] have suggested how site-directed mutagenesis of a given pair of immunoglobulin Heavy- and Light-chain genes, together with functional analysis of the proteins in an appropriate expression system and computer graphic modelling, could be used to explore the structure/function relationships of the antigen-combining sites of antibodies (Fig. 1). From such approaches it is possible to conceive how empirical rules for structure prediction of variable-region domains of antibodies could be devised and refined. Such an approach could be extended to other proteins where similar structural information and functional assays become available.

3.4.4 Quality Control in Production of Recombinant Proteins

An increasing number of recombinant polypeptide structures with specific biological activities have also been generated for therapeutic application in man. These include

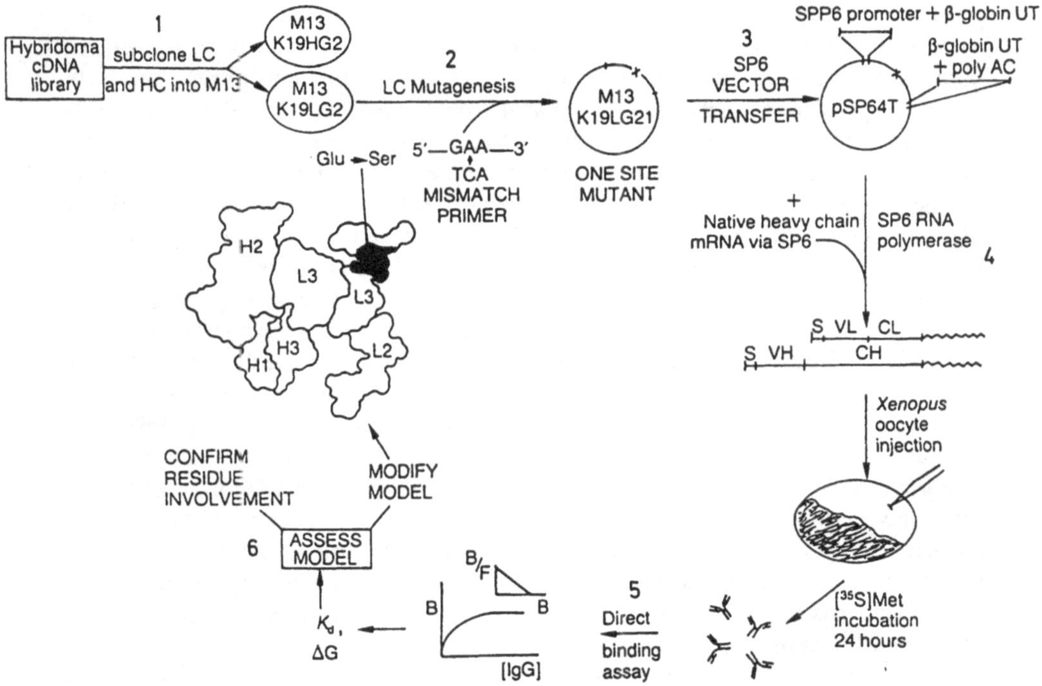

Fig. 1. Probing the structure of an antigen-combining site of a monoclonal antibody using site-directed mutagenesis. The figure shows a scheme, taken from Rees and de la Paz (1986) [36], illustrating a hypothetical experimental cycle devised to determine the relationship between the structure of an antibody light-chain and its antigen-combining properties. A cDNA library is constructed using template mRNA isolated from the hybridoma which generates the monoclonal antibody, from which heavy chain (HC) and light chain (LC) immunoglobulin cDNAs are identified (*1*). In oder to study the functional consequences of incorporating a specific change in the LC sequence (Glu to Ser), the LC cDNA is subjected to site-directed mutagenesis using a mutagenic oligonucleotide, which creates a mismatch in the coding sequence when incorporated into DNA and replicated. This step can be performed either with the aid of M13 bacteriophage vectors (as illustrated) or with alternative vectors used for this purpose (*2*). The mutant cDNA, after sequence analysis, is incorporated into an SP6 vector with appropriate transcriptional and translational control signals to permit the in vitro synthesis of functional mutant cDNA (*3*). The mutant LC mRNA is then injected into *Xenopus* oocytes, together with normal HC mRNA prepared in a similar manner in vitro (*4*). Radioactively-labelled antibody isolated from the oocytes is used in antigen-binding assays to determine the effect(s) of the LC mutation (*5*). This either confirms, or otherwise, the involvement of the specific residue in question and three-dimensional computer model building can be used to refine the empirical rules governing antibody-antigen recognition (*6*). Similar experiments could be used to probe the structure/function of the HC sequence. The scheme could be extended to include 3D-structural analysis of antibody Fab fragments using X-ray diffraction analysis or an alternative procedure. (Reproduced with permission of the original Publishers, Elsevier Press Ltd.)

recombinant vaccines, hormones and even larger protein structures: human insulin, somatostatin, β-urogastrone, human growth hormone, tissue plasminogen activator, human interferons and recombinant antibodies are but a few examples. In the majority of cases these molecules are produced in bioreactors using either bacteria, yeast or animal-cells to express the required recombinant protein.

The production of these molecules as therapeutics using these new production systems has necessitated the introduction of appropriate procedures for quality control, purification and toxicological testing of the recombinant protein products, since the possibility exists that they may be contaminated with cellular impurites such as other proteins, polysaccharides, DNA or possibly viruses, derived from the bioreactor. Regulatory authorities therefore require detailed analytical information on the degree of purity of recombinant proteins that are destined for use as in vivo diagnostics or as therapeutics in man. Regulations for quality control in the production of recombinant proteins are now reasonably well established and are being continuously updated [37–39]. For smaller protein molecules, amino acid analysis and peptide mapping may provide adequate analytical evidence of purity. Reverse phase HPLC can be used effectively for large proteins such as antibodies [40], and methods such as circular dichroism can provide useful additional analytical information which correlates not just with product purity, but also with biological activity, to take beta-urogastrone as an example [41]. Additional standard protein analytical methods are also used to assess product purity, including SDS-PAGE, isoelectric focusing, ELISA tests, and immunoblotting for minor protein contaminants (reviewed by Thomas [42]). Nucleic acid hybridization techniques are used to estimate the level of DNA contamination [43,44], and viruses are detected using a combination of infectivity tests and specific assays for virus-specific enzymes [39]. Finally, the product must be shown to contain tolerably low levels of pyrogens/endotoxins [45].

In the bioreactor itself it is necessary to demonstrate that altered forms of the recombinant product do not arise as a result of spontaneous mutation. Hence, it is necessary to confirm the sequence of the recombinant protein and/or gene for each batch culture, or in samples taken periodically from continuous cultures. The required methodologies for the rapid determination of the primary nucleotide sequence/polypeptide sequence of the gene and protein have already been developed and automated. However, fast and routine methods for obtaining three-dimensional structural information could also be useful as a means to demonstrate structural identity (or reveal structural dissimilarities) between recombinant and natural protein, and to detect impurities or contaminating mutant forms arising in the bioreactor. In this regard the NMR method, because of its speed and applicability to a wide-range of globular proteins, could have considerable potential as a future analytical tool for bioprocess control.

4 Sensors, Biosensors and Their Applications

4.1 Concept and Applications

Sensor and biosensor devices have a wide range of potential applications, particularly in the human and veterinary health care industries. They were conceived as a result of a medical need for rapid analysis of specific substances as diagnostic indicators of disease, or for continuous on-line in vivo monitoring of critical body functions. The first medical biosensors were produced in the United States and although novel, these early devices were often cumbersome and somewhat limited in their application.

Modern biosensors benefit from the considerable technological innovation that has taken place in this field during the past decade.

The medical care industry is still one of the major markets for the application of biosensors. Current estimates of the world market for medical diagnostics are in the region of $ 4,000 million [47]. Interestingly, in a recent review containing a list of the current 24 major producers of medical and commercial biosensors twelve were located in Japan, seven in the U.S.A., one in Canada, one in Eastern Europe, one in Scandinavia, but only two in Western Europe [47].

With regard to other applications, there is an increasing demand worldwide for specific and sensitive methods for monitoring levels of toxic substances, heavy metal ions and pesticides in the environment. Likewise modern and highly-mechanized food production processes require reliable and sensitive methods to detect microbial contamination and to monitor concentrations of nutrients and various food additives. Finally, there are several potential applications of biosensors in bioprocess control, for monitoring biomass, the rate of utilization of substrates and the production of the microbial product.

4.2 Structure of Biosensors

4.2.1 General Aspects

In most instances biosensors have been adapted from simpler devices designed originally for the measurement of relatively simple analytes: Oxygen, CO_2, pH, sodium, potassium or calcium, NH_3. Although limited in their analytical potential, such sensors provide extremely valuable information for the clinician wishing to monitor patient condition continuously in a critical situation, for example following heart attack or in a patient receiving kidney dialysis.

The basic concept behind the construction of a biosensor is shown schematically in Fig. 2 [46]. Biosensors differ from simpler gas, vapour and ion sensors in that they have a mechanism to detect selectively any given substance in a complex matrix containing a mixture of other components. The selectivity of the device for the complex analyte is made possible by molecular recognition using a biological component such as an enzyme, antibody, cell membrane or whole cell located in the biosensor matrix. A signal is thus generated as a result of the interaction which is then transduced to an appropriate detector. The recognition event is designed either to produce/consume a simple secondary molecule whose level is monitored using a gas- or ion-sensitive sensor [47], or to induce some physical change which can be detected by an optical or electronic device.

All biosensors share this basic design concept, and ingenious solutions to problems of biosensor construction have arisen from the interface between electronic engineering, biochemistry/enzymology and molecular biology. For new medical applications current biosensor technology is being focused on a few central issues: Ideally medical biosensors should be low-cost, sterilizable and disposable, give reproducible results, and in many cases be suitably miniaturized for on-line monitoring of patient condition. Such miniaturized devices avoid the trauma associated with the introduction of larger, more impractical monitoring instruments into the blood vessels of a patient.

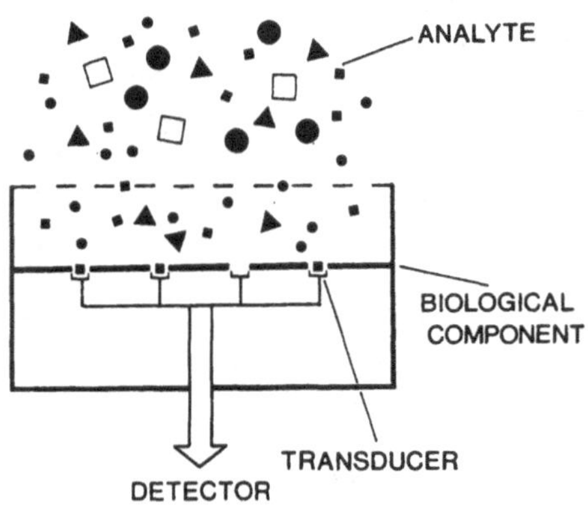

Fig. 2. General concept underlying the contruction of a biosensor. Adapted from Rolfe and Martin (1988) [46]. A specific analyte (■) present in a complex matrix containing other substances (other symbols) is recognised by the biological component of the biosensor. The biological component may be an enzyme, an antibody, or another material with which the analyte specifically interacts. As a result of the interaction a secondary signal is generated within the biosensor which is transduced to a detector via the transducer component. Interfering substances in the complex matrix can sometimes be selectively excluded from interaction with the biological component by incorporation of selective membranes into the biosensor (*broken line*)

Some of these design features are also required of biosensors used for on-line monitoring of bioreactors. The major problem to be overcome is that such biosensors should be resistant to heat and chemical sterilization procedures so that microbial or fungal contamination can be avoided when the instrument is introduced into the production vessel.

4.2.2 Biological Components

Enzymes have been widely used as the analyte-selective components of biosensors. However, the development of enzyme-based biosensors has been hampered because of problems with enzyme stability, or where complex cofactors are required for enzymic activity. The former problem may be circumvented by using thermostable enzymes, either obtained from natural sources such as thermophylic microorganisms, or by using recombinant genetics to engineer heat-stable variants of otherwise thermosensitive enzymes (Sect. 3.4).

Multienzyme complexes located in situ in plant or animal tissues have also been used in biosensor construction, avoiding the need to undertake costly steps to purify enzymes or supply cofactors. Moreover, where whole organisms are used as the biological component, the organisms can be regenerated by immersion of the biosensor in a suitable growth medium. Selectivity of such complex biosensors is also amenable to manipulation by generating mutants, either by conventional means or by recombinant genetics, in order to identify and eliminate interfering metabolic pathways [48,49]. Possibly because the active enzyme species are in a more native environment within a tissue or cell such complex biosensors can be quite stable. Additionally, they often display a remarkably high level of sensitivity and have surprisingly fast response times compared with biosensors which incorporate purified enzymes [50]. In contrast, biosensors containing purified subcellular organelles or cell membranes have so far proved to be less stable and more costly to prepare [51]. Receptors, because of their natural high specific affinity for their biological ligands, have also recently been considered as good candidates for biological components of biosensors [52]. Major obstacles are that natural receptors are often not available in useful quantities, they have a short half-life in biosensors, and they are often optimally active only when maintained in vesicles [47]. Molecular cloning of receptor genes, and specific modification of their structure to generate engineered molecules suitable for use in biosensors may overcome at least some of these problems and limitations in the future (Sect. 3.3).

Apart from the choice of biological component, another important issue is the choice of technique used to immobilize the enzyme or tissue within the biosensor matrix. Many alternative techniques are now available including adsorption, cross-linking, covalent binding, entrapment or encapsulation [53]. All these procedures have some advantages and disadvantages; the choice of method depends largely on the particular application and is determined empirically.

4.2.3 Transducers

Once the specific analyte is detected by the biological component of the biosensor the signal is converted to an electronic or visible signal by the transducer, after which it is directed to an appropriate detector (Fig. 2). Various types of transducer have been developed for, or adapted to, application in biosensors.

Electrochemical Transducers

Ions, or simple substances such as gases, generated by the action of the enzymic component of the biosensor, can be detected by electrochemical transducers. Potentiometric transducers of the ion-selective type (pH, fluoride, NH_4^+) and gas-sensitive type (CO_2, NH_3) have been used in biosensors. Ion-sensitive transducers suffer from interference with other ions, whereas gas sensors in general have slower response and recovery times [54]. Amperometric oxygen and peroxide electrodes have likewise been used as biosensor transducers [55] and have similar limitations. H_2O_2 electrodes, for example, are subject to interference by compounds such as glutathione, ascorbic acid or uric acid. Appropriate selectively-permeable membranes can, in some instances, be incorporated to eliminate such interference [55]. The versatility and reliability of electrochemical transducers is attested by the fact that they have been incorporated into biosensors for over 80 different substances [47].

New approaches to the construction of electrochemical transducers, by adapting microelectronic techniques, has led to a number of novel microfabricated biosensors, such as the glucose oxidase-based glucose sensor developed by de Rooij and collaborators [56]. Such biosensors, due to their small size and simple construction, have a special place in critical health care as disposable, low cost devices for continuous in vivo monitoring of body function.

Calorimetric Transducers

Since most enzyme-catalyzed chemical reactions are exothermic, heat-responsive transducers can also be used as sensitive devices to monitor the signal generated by the enzyme component of a biosensor. They are able to detect enzyme-specific substrates at concentrations as low as 10^{-5} M. Calorimetric biosensors have been devised for the detection of many substances, ranging from sugars (glucose, sucrose), ethanol, antibiotics (Gentamycin, penicillins), proteins and hormones (albumin, insulin) and various other substances and metabolites (ascorbic acid, ATP, lactate, oxalic acid, triglycerides, urea, uric acid).

Since, without exception, all growth processes are exothermic, calorimetric measurements are also increasingly being recognised as providing important information on the metabolic state of a microbial cell culture, for on-line control of bioreactors [57–58].

Other Transducers

Field effect transistors (FET) can also be used to monitor the charge build up on an ion-sensitive electrode resulting from interaction of the biological component of the biosensor with the analyte. The combination of a thin layer of enzyme gel over an ion-sensitive membrane placed directly over the gate of an FET allows the device to monitor a specific substance, for example glucose using glucose oxidase [59], ATP using ATPase [60], or ethanol using alcohol oxidase [61].

Similar new developments have taken place in the application of piezoelectric (PE) crystals as transducers in immunological biosensors. This technology uses the principle that the frequency of oscillation of the PE crystal is altered when its effective mass changes due to binding of antigen to a specific antibody previously deposited on its surface. Karube and Gotoh [62] have used specific antibody directed against a surface antigen of *Candida albicans* to construct a PE biosensor specific for this organism. Surface acoustic wave detectors have been used in an analogous manner [63]. Optical immunobiosensors are also attracting considerable attention, since they can exploit the high specificity of antibody molecules as the biological component in combination with the high sensitivity of fluorescent detection technology. Microfibre optic technology is being employed in the structure of miniaturized versions of these devices [46].

4.3 DNA Sensors

Apart from conventional biosensors, new developments are also taking place in DNA sensor technology. Sensitive DNA sensors capable of rapid analysis will have an important application in Biotechnology process development, in particular as devices to replace existing methods for quantitative monitoring the level of DNA contamination or detection of virus-specific nucleic acids in recombinant protein preparations [39,43,44] (see Sect. 3.4.4).

One type of DNA biosensor already constructed along conventional lines uses two DNA binding proteins, one to link DNA to a membrane within the biosensor, a second to label the DNA, and a light-emitting diode to monitor the response. The device is reported to permit the detection of 2–1,000 pg of DNA in a protein solution [64], which is in a range which would be useful to replace current hybridization methods [43].

A more novel approach to detect specific DNA sequences is to use an in vitro DNA amplification system using a "polymerase chain reaction" (PCR), the principle of which is illustrated in Fig. 3. The technology uses a thermostable DNA polymerase (currently from *Thermus aquaticus*) in combination with specific oligonucleotide primers and appropriate nucleotides, to amplify specific nucleotide sequences [65]. The temperature-shifts required to undergo successive rounds of DNA denaturation,

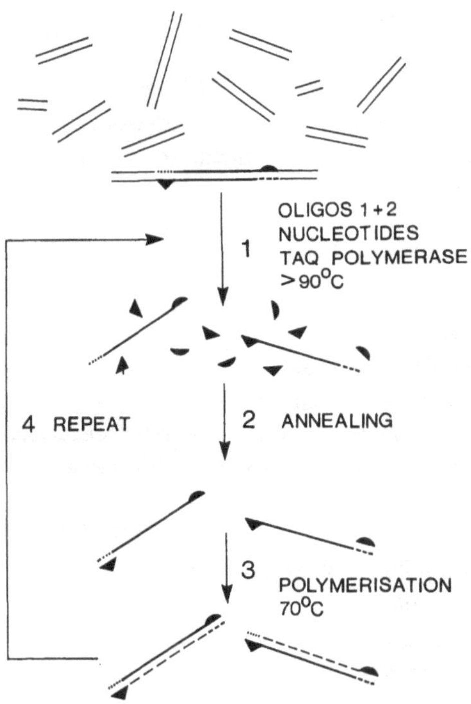

Fig. 3. Principle of in vitro amplification of DNA using the polymerase chain reaction (PCR). In the example illustrated, PCR is used to amplify specifically a genomic DNA segment located between two short, defined nucleotide sequences (thick duplexes, bounded by regions marked . . . . and - - - -) present in a mixture of other DNA fragments (thin duplexes). The DNA fragments are denatured at high temperature (>90 °C) in the presence of two synthetic oligonucleotides designed to be complementary to the ends of the target sequence to be amplified (*1* and *2*; ▲ and ▲) together with thermostable (Taq) DNA polymerase and oligonucleotides (*1*). After annealing of the oligonucleotides to the complementary DNA sequences on the template strands under appropriate conditions (*2*), the temperature is shifted and the polymerase synthesises the specific complementary DNA strands from each primer (*3*). The products of the reaction are DNA duplexes in which the oligonucleotide annealing sites at their termini are duplicated. After denaturation, the reaction products are capable of undergoing further rounds of specific amplification (*4*). Each successive cycle leads to a theoretical doubling in the amount of amplified sequence. Selective amplification depends on careful choice of primer oligonucleotides and amplification conditions. In practice successive cycles of denaturation, annealing and DNA chain elongation are carried out using an automated, programmable heating device. Together with a suitable DNA detector system, such a procedure can be used to monitor the presence of minute amounts of any specific DNA sequence, such as from a virus, plasmid or bacterium

annealing and polymerization are carried out in automated heating devices. The method is sensitive enough to amplify a specific sequence from a single sperm or cell [66]. Although still in the early stages of development several envisaged applications of this technology can be foreseen because of its exquisite specificity and sensitivity, both in the diagnostics industry and for monitoring recombinant products and foodstuffs for virus and microbial contamination.

Acknowledgements: The author is grateful to participants of the First Swiss-Japanese Joint Meeting on Bioprocess Development [1] for making information available for the preparation of this article, to colleagues for helpful discussions, and to Dr. A. Fiechter and Dr. B. Sonnleitner, ETH Zurich, for their helpful comments during preparation of the manuscript.

5 Addendum

1st Swiss-Japanese Joint Meeting on Bioprocess Development

Opening lecture: S. Fukui

Protein structure. Structure and function of enzymes; chaired by: **K. Wüthrich**

K. Wüthrich	NMR techniques for the characterization of proteins in solution
J. Jansonius	Protein structure determination by X-ray diffraction of single crystals
H. Okada	Xylanase of *Bacillus pumilus*; its tertiary structure and estimation of the catalytic site
H. Nakamura	Calculations of electrostatic aspects of proteins

Protein design. Bioregulation of enzyme activity; chaired by: **H. Yamada**

R. Ghosh	Membrane protein engineering in photosynthetic prokaryotes
H. Tanaka	Synthesis of selenium analogues of metallothionein from *Neurospora crassa* and their properties

Analytics and process control; chaired by: **A. Fiechter**

M. Grätzel	Molecular and bioelectronics
U. von Stockar	On line calorimetry for process control
M. Widmer	Toward on line application of biosensors
N. de Rooij	FET for applications in biochemical engineering
K. Soda	Application of amino acid racemases: synthesis of stereoselectively 4-labeled NAD^2H and stereospecificity analysis of H transfer of NADH

Enzyme synthesis and purification (inc. large scale); chaired by: **T. Beppu**

C. Leist	t-PA synthesis with human cells
H. Kocher	Chimeric monoclonal antibodies
H. Walliser	Large-scale purification of proteins
E. Hochuli	Purification of interferons
Y. Yokoo	Correctly refolded salmon growth hormone derived from *E. coli*
T. Beppu	Site directed mutagenesis of the milk-clotting aspartyl protease, chymosin and *Mucor* rennin

New function of enzymes and their application; chaired by **U. von Stockar**

H. Kumagai	γ-Glutamyl transpeptidase from *E. coli.* Gene cloning, rapid purification and synthesis of γ-glutamyl compounds
K. Yokozeki	Enzymatic production of ribavirin from natural nucleosides
M. Grütter	Protein modelling
A. Tanaka	Yeast peroxisomes as organized assemblies of specific enzymes
H. Yamada	Enzymatic processes for the production of nicotinamide and nicotinic acid with *Rhodococcus rhodochrous* J1 cells

General discussion; Closing

A. Fiechter Conclusions
H. Ursprung Japanese-Swiss realtionships

6 References

1. Fiechter A (1988) New Frontiers of Bioprocess Development: First Swiss-Japanese Meeting on Bioprocess Development, Interlaken, Switzerland, 25–27 July 1988 (s. Addendum, this work).
2. Asano Y, Tachibana M, Tani Y, Yamada H (1982) Agric. Biol. Chem. *46*: 1175
3. Nagasawa T, Nanba H, Ryuno K, Takenkhi K, Yamada H (1987) Eur. J. Biochem. *162*: 691
4. Nagasawa T, Ryuno K, Yamada H (1986) Biochem. Biophys. Res. Comm. *139*: 1305
5. Nagasawa T, Mathew CD, Manger J, Yamada H (1989) Appl. Environ. Microbiol. (in press)
6. Yamashita T, Tonouchi N, Uozumi T, Beppu T (1987) Mol. Gen. Genetics *210*: 462
7. Suzuki H, Humagai T, Echigo T, Tochikura T (1988) Biochem. Biophys. Res. Comm. *150*: 33
8. Shirae H, Yokozeki K, Uchiyama M, Kubata K (1988) Agric. Biol. Chem. *52*: 1777
9. Rink H, Liersch M, Sieber P, Meyer F (1984) Nucleic Acids. Res. *12*: 6369
10. Bode W, Papamokos E, Djordje M, Seemueller U, Fritz H (1986) EMBO J *5*: 813
11. Kline AD, Brown W, Wuethrich K (1986) J. Mol. Biol. *189*: 377
12. Wagner G, Braun W, Havel TF, Schaumann T, Go N, Wuetherich K (1987) J. Mol. Biol. *196*: 611
13. Clore GM, Gronenborn AM, James MNG, Kjaer M, McPhalen CA, Poulson FM (1987) Protein Eng. *1*: 313
14. Braun W, Wider G, Lee KH, Wuethrich K (1983) J. Mol. Biol. *169*: 921
15. Vasak M, Worgotter E, Wagner G, Kagi JHR, Wuethrich K (1987) J. Mol. Biol. *196*: 711
16. Wuethrich K (1986) NMR of proteins and nucleic acids. Wiley, New York
17. Blank U, Ra C, Miller L, White K, Metzger H, Kinet J-P (1989) Nature *337*: 187
18. Davis MM, Bjorkman PJ (1988) Nature *334*: 395
19. Navia MA, Fitzgerald PMD, McKeever BM, Leu C-T, Heimbach JC, Herber WK, Signal IS, Darke PL, Springer JP (1989) Nature *337*: 615
20. Wagner E, Covarrubias L, Stewart TA, Mintz B (1983) Cell *35*: 647
21. Kurt-Jones EA, Hamberg S, Ohara J, Paul WE, Abbas AK (1987) J. Exp. Med. *166*: 1774
22. Pouwels PH, Enger-Valk BE, Brammar WJ (1985) Cloning vectors. Elsevier, Amsterdam
23. Heinz DW, Liersch M, Braun NJ, Gruetter M (1988) New inibitory properties of Eglin-C after specific mutagenesis. In: Wittman-Liebold (ed) Methods in Protein Sequence Analysis: Proc. 7th Int. Conf. Protein Sequence Analysis (in press)
24. Leu FT-K, Fersht AR (1987) Nature *326*: 811
25. Blundell TL, Sibanda BL, Sternberg MJE, Thornton JM (1987) Nature *326*: 347
26. Jones TA, Thirup S (1986) EMBO J. *5*: 819
27. Protein Engineering Research Institute (PERI). Osaka, Japan, Director: Dr. M. Ikehara
28. Morrison SL (1985) Science *229*: 1202
29. Nishimura Y, Yokoyama M, Araki K, Ueda K, Ueda R, Kudo A. Watanabe T (1988) Cancer Res. *47*: 999
30. Sun LK, Curtis P, Rakowicz-Szulczyunska E, Ghrayeb J, Chang N, Morrison SL, Koprowski H (1984) Proc. Nat. Acad. Sci. U.S.A. *84*: 214
31. Williams GT, Neuberger MS (1986) Gene *43*: 319
32. Gascoigne NRJ, Goodnow CC, Dudzik KL, Oi VT, Davis MM (1987) Proc. Nat. Acad. Sci. U.S.A. *84*: 2936
33. Jones PT, Dear PH, Foote J, Neuberger MS, Winter G (1986) Nature *321*: 522
34. Lerner RA, Tramontano A (1987) Trends in Biol. Sci. *12*: 427
35. Napper AD, Benkovic SJ, Tramontano A, Lerner RA (1987) Science *237*: 1041
36. Rees AR, de la Paz P (1986) Trends in Biol. Sci. *11*: 144
37. WHO Expert Committee on Biological Standardisation (1987) WHO Technical Report Series, WHO, Geneva, p 93
38. Ad hoc Working Party on Biotechnology/Pharmacy (1987) Trends in Biotechnol. *5*: G1

39. Ad hoc Working Party on Biotechnology/Pharmacy (1988) Trends in Biotechnol. *6*: G5
40. Bigee WL, Vanderlaan M, Fong SSN, Jensen RH (1983) Mol. Immunol. *20*: 1353
41. Brewer SJ (1986) J. Chem. Technol. Biotechnol. *36*: 367
42. Thomas AH (1988) Chemistry in Britain *24*: 1031
43. WHO Study Group (1987) WHO Technical Report Series 747, WHO, Geneva, p 1
44. Goldfarb P (1986) J. Chem. Technol. Biotechnol. *36*: 389
45. Dinarello CA (1974). In: Adams D, Edelson P, Koren H (eds) Methods for studying mononuclear phagocytes. Academic, New York, p 69
46. Rolfe P, Martin MJ (1988) Chemistry in Britain *24*: 1026
47. Luong JHT, Mulchandani A, Guilbault GG (1988) Trends in Biotechnol. *6*: 310
48. Rechnitz GA (1978) Chem. Eng. News *56*: 56
49. Arnold MA, Rechnitz GA (1981) Anal. Chem. *53*: 515
50. Arnold MA, Rechnitz GA (1987) In: Turner APF, Karube I, Wilson GS (eds) Biosensors: Fundamentals and applications. Oxford University Press, Oxford, p 30
51. Burstein C, Adamowicz E, Boucherit K, Rabouille C, Romette JL (1986) App. Biochem. Biotechnol. *12*: 1
52. Wingard LB Jr (1987). In: Schmid RD, Guilbault GG, Karube I, Schmidt H-L, Wingard LB Jr (eds) Biosensors: International Workshop, GBF Monographs, Braunschweig, vol 10 p 133
53. Coughlin MP, Kiertan MPJ, Border PM, Turner APF (1988) J. Microb. Methods *8*: 1
54. Kuan SS, Guilbault GG (1987). In: Turner APF, Karube I, Wilson GS (eds) Biosensors: Fundamentals and applications. Oxford University Press, Oxford, p 135
55. Clark LC Jr (1979) Methods Enzymol. *44*: 2161
56. Koundelka M, Gernet S, de Rooij NF (1987) Voltommetry, a powerful tool for evaluation and process control of thin film electrodes. In: Transducers '87, Tokyo, p 41
57. Marison IW, von Stockar U (1987) Enzyme and Microb. Technol. *9*: 33
58. Birou B, Marison IW, von Stockar U (1987) Biotechnol. Eng. *30*: 650
59. Hanazoto Y, Shiono S (1983). In: Seiyama T (ed) Proc. Int. Meeting on Chemical Sensors, Fukuoka. Anal. Chem. Symp. Series, Elsevier, Amsterdam, vol 17 p 523
60. Adanielsson B, Lundstrom F, Winquist F, Mosbach K (1979) Anal. Lett. *B12*: 1189
61. Karube I, Kubo I (1987). In: Guilbault GG, Mascini M (eds) Analytical uses of immobilised biological compounds for detection, medical and industrial uses. Reidel, p 207
62. Karube I, Gotoh M (1987). In: Guilbault GG, Mascini M (eds) Analytical uses of immobilised biological compounds for detection, medical and industrial uses. Reidel, p 267
63. Roederer E, Baastians GJ (1983) Anal. Biochem. *55*: 2333
64. Knight P (1989) Biotechnology *7*: 175
65. Saiki RK, Gelford DH, Stoffel S, Scharf SJ, Higuchi R, Horn GT, Mullis KB, Erlich HA (1988) Science *239*: 487
66. Li H, Gyllensten UB, Cui X, Saiki RK, Erlich HA, Arnheim N (1988) Nature *335*: 414

Protein Purification

Erik P. Lillehoj
Biotech Research Laboratories, Inc., 1600 East Gude Drive, Rockville, MD 20850, USA

V. S. Malik
Philip Morris Research Center, P. O. Box 26583, Richmond, VA 23261, USA

Abbreviation ... 20
1 Introduction-Historical Perspective ... 20
2 Development of a Protein Purification Strategy ... 22
2.1 General Considerations ... 22
2.2 Sample and Methods Characteristics Influencing Purification ... 22
2.3 Assessment of Purity ... 24
3 Choice of Starting Material ... 26
3.1 Relative Protein Abundance ... 26
3.2 Physical State of the Protein ... 27
4 Purification of Proteins in the Denatured State ... 30
4.1 Protein Native State ... 30
4.2 Denaturation-Renaturation ... 30
4.3 Inclusion Bodies ... 31
5 Preparation of Cell-Free Extracts ... 33
6 Bulk Precipitation and Phase Extraction ... 33
7 Conventional vs High-Performance Liquid Chromatography ... 34
7.1 The Principle of Liquid Chromatography ... 34
7.2 High-Performance Liquid Chromatography ... 34
7.3 Mathematical Descriptions ... 36
8 Descriptions of Chromatographic Techniques ... 37
8.1 Size Exclusion Chromatography ... 38
8.2 Ion Exchange Chromatography ... 41
8.3 Hydroxylapatite Chromatography ... 44
8.4 Reversed Phase Chromatography ... 44
8.5 Hydrophobic Interaction Chromatography ... 47
8.6 Affinity Chromatography ... 48
9 Optimizing Preparative-Scale Isolations ... 52
10 Descriptions of Electrophoretic Techniques ... 53
10.1 Sodium Dodecyl Sulfate Polyacrylamide Gel Electrophoresis ... 54
10.2 Isotachophoresis ... 56
10.3 Isoelectric Focusing ... 57
10.4 Two-Dimensional Polyacrylamide Gel Electrophoresis ... 58
11 Amino Acid Sequencing of Proteins ... 59
11.1 Micro-Sequence Analysis ... 59
11.2 Radiochemical Sequence Analysis ... 61
12 Recombinant DNA in Protein Overproduction ... 61
12.1 Prokaryotic Expression ... 61
12.2 Eukaryotic Expression ... 61
13 Fusion Proteins ... 62
13.1 β-Galactosidase Fusion Proteins ... 62
13.2 Other Fusion Proteins ... 62
14 Future Developments ... 64
15 References ... 65

This monograph summarizes recent developments in the purification and analysis of natural and recombinant proteins. The basic strategies employed in protein purification are reviewed with regards to the characteristics of the protein of interest that may aid its isolation, choice of the starting material, and use of denaturants. Preparation of cell-free extracts followed by bulk precipitation and/or phase partition constitute the initial steps of many purification schemes. Chromatographic methods (size exclusion, ion exchange, hydroxylapatite, reversed phase, hydrophobic interaction and affinity based) utilizing either traditional, low pressure or high-performance liquid chromatography instrumentation are discussed. Electrophoretic techniques used to analyze the homogeneity of the protein product include SDS-PAGE, isotachophoresis, IEF and two dimensional gel electrophoresis.

Abbreviations

a, selectivity factor; ABx, antibody exchanger; cDNA, complementary DNA; CM, carboxymethyl; CoA, coenzyme A; D, diameter; Da, Daltons; DEAE, diethylaminoethyl; DNA, deoxyribonucleic acid; FPLC, fast protein liquid chromatography; FR, flow rate; HETP, height equivalent to a theoretical plate; HIV-1, human immunodeficiency virus type 1; HPLC, high-performance liquid chromatography; IEF, isoelectric focusing; IgG, immunoglobulin G; IgM, immunoglobulin M; K, distribution coefficient; k′, capacity factor; L, length; LV, linear velocity; N, number of theoretical plates; PAGE, polyacrylamide gel electrophoresis; PEI, polyethyleneimine; PEG, polyethylene glycol; pI, isoelectric point; PMSF, phenylmethylsulfonyl fluoride; QAE, quaternary aminoethyl; R, resolution; RNA, ribonucleic acid; Rt, retention time; SDS, sodium dodecyl sulfate; SF, scale-up factor; SP, sulfoxypropyl; TFA, trifluoroacetic acid; V, retention volume; V_0, void volume.

1 Introduction-Historical Perspective

Early methods of protein separations date to the beginning of the current century. Analytical separations using electrophoretic techniques were developed shortly after electric power became commercially available. These were simply performed in liquid media filling a U-shaped glass tube with the electrodes suspended in each of the open ends. Early methods of preparative-scale protein isolations relied upon their solubility characteristics. Prior to the 1940's, acidic, organic and salt solutions were used to precipitate proteins from complex mixtures of biological fluids. However, as solubility-based purification methods were refined, it became increasingly apparent that not only were these relatively crude in their ability to produce a homogeneous protein preparation but also often produced a preparation devoid of biological activity. During the 1940's, ultracentrifugation appeared as a powerful new tool that offered the ability to separate proteins on the basis of size rather than solubility characteristics. Although biological activity was often retained after such a separation, it was several years before ultracentrifugation was applied to preparative protein isolations. Electrophoretic methods of protein separations were also improved during this period. Liquid media utilized in earlier times were replaced by various types of solid supports (e.g. paper, starch, agar and polyacrylamide). These offered advantageous properties of mechanical rigidity as well as chemical inertness that improved analytical resolution of protein mixtures.

The next two decades witnessed the development of two new chromatographic techniques that revolutionized the field of protein chemistry. Initially, ion exchange and later size exclusion chromatography were introduced as novel, preparative-scale techniques based upon protein charge and size, respectively. Methods of electrophoresis were also refined, by the use of discontinuous pH conditions and addition of a variety of modifying agents, particularly detergents (e.g. sodium dodecyl sulfate). The utility that both of the new chromatographic and electrophoretic methods offered to the protein chemistry laboratory is verified by their continued popularity today. Two important developments during the 1960's and 1970's greatly enhanced the field of protein purification. Affinity chromatography, appearing in the latter part of the 1960's, allowed the isolation of proteins based on their biological rather than physicochemical characteristics. This was a particularly important development since the number of different physical or chemical properties upon which to base separations is limited. It is not uncommon to achieve an operationally complete purification of a protein from crude starting material using a single affinity-based separation protocol as, for example, the purification of immunoglobulin G from serum using *Staphylococcus aureus* protein A bound to a solid support [1]. The second development, high-performance liquid chromatography (HPLC), introduced in the late 1970's, has not only led to second generation application of established techniques of conventional liquid chromatography (ion exchange, size exclusion and affinity chromatography) but also the direct development of new technologies (reversed phase and hydrophobic interaction chromatography). The future of protein separations will continue to rely on the application of new products (e.g. fluoropolymers) and developments to chromatography and electrophoresis such as radial flow, supercritical fluid and open tubular liquid chromatography and capillary electrophoresis.

The following monograph is an attempt to summarize the state of the art of protein purification. We have directed this review primarily toward those individuals who are new to the field and desire a concise treatise from which they can refer to the original literature for potential application to their own problem. Although we have attempted to summarize as many aspects of protein purification as possible, the reader is advised that it is impractical to cover such a broad topic (on which entire texts have been written) without inadvertently omitting some areas. Thus, the references cited herein are intended only to serve as examples and should not be considered as inclusive. In most instances, these references have been limited to the past four to five years to provide the reader with the most up-to-date developments. Other extensive reviews have recently appeared and may be consulted should further assistance be necessary. Several books have been published that deal exclusively with both theoretical and practical aspects of protein purification [2–5]. Others have appeared in the scientific literature as compilations [6] or general review articles [7–18] or specifically limited to the purification of selected proteins such as peptide hormones [19], growth factors [20], enzymes [21,22] or proteins of plant [23,24], viral [25] or membrane [26] origin. Specific types of separation methods have also been summarized including size exclusion [27,28], ion exchange [29], hydrophobic interaction [30] and affinity [31–34] chromatography.

2 Development of a Protein Purification Strategy

2.1 General Considerations

Three basic steps are performed when attempting to isolate a protein. First, small-scale, analytical experiments are used to analyze the feasibility of a particular separation technique on the protein of interest. These are organized in a sequence such that the end product is free of contaminating proteins. Second, the process is increased to a preparative-scale allowing sufficient material to be isolated at the end of the process. Finally, the purity of the isolated protein is checked. Often it may be necessary to amend the purification scheme based upon the final degree of purity since contaminants may appear during the large-scale purification that were not detected during the small-scale experiments. In this case it is convenient to analyze a small amount of material from the last stage of the preparative-scale separation scheme, scale up to the preparative mode if satisfactory and reexamine for purity. Thus, one can readily appreciate the trial-and-error process that must usually be performed to achieve an end product of acceptable purity. It is important to emphasize that the concept of "acceptable purity" varies with the protein at hand. For example, the purity of a protein with demonstrable biological activity such as an enzyme may simply mean freedom from inhibitors or interfering substances. On the other hand, purification of structural proteins often requires evidence of the absence of all detectable contaminants.

2.2 Sample and Methods Characteristics Influencing Purification

There are several factors to be considered when the initial small-scale purification strategy is conceived. Foremost among these is an understanding of the biological and physicochemical properties of the protein one is attempting to isolate. Does it possess a highly specific ligand that could be exploited in an affinity-based separation protocol? Do its charge, molecular weight, isoelectric point and/or hydrophobicity lend themselves to a particular isolation technique that should be undertaken early in the purification scheme? A second factor to consider is the objective for purification. To retain biological activity it is normally required that all separation techniques be compatible with maintaining secondary, tertiary and quaternary structure. Limitation to nondenaturing conditions to preserve protein conformation restricts one's choice of techniques that can be employed. On the contrary, if biological activity needs not be preserved, as for example when the objective is to prepare a protein suitable for amino acid sequence analysis, the list of potential separation techniques includes those that are denaturing as well as nondenaturing.

Independence of the methods chosen also affects the efficacy of a particular purification stategy. A common practice is to sequentially use separation techniques founded upon different physicochemical properties and which maximize the greatest differences between the physical properties of the protein of interest and its contaminants. In this regard, affinity chromatography, if available, should be considered in every purification scheme since it is the only commonly available technique that is based on the biological rather than physical properties of proteins. Affinity-based separation methods that rely on the antigenic characteristics of proteins should be considered in the same light. Finally, the actual sequence of steps one chooses may greatly affect the final yield and purity of the end product. It is a general rule of thumb to utilize a different mode of separation at each succeeding step [35]. Figure 1 illustrates

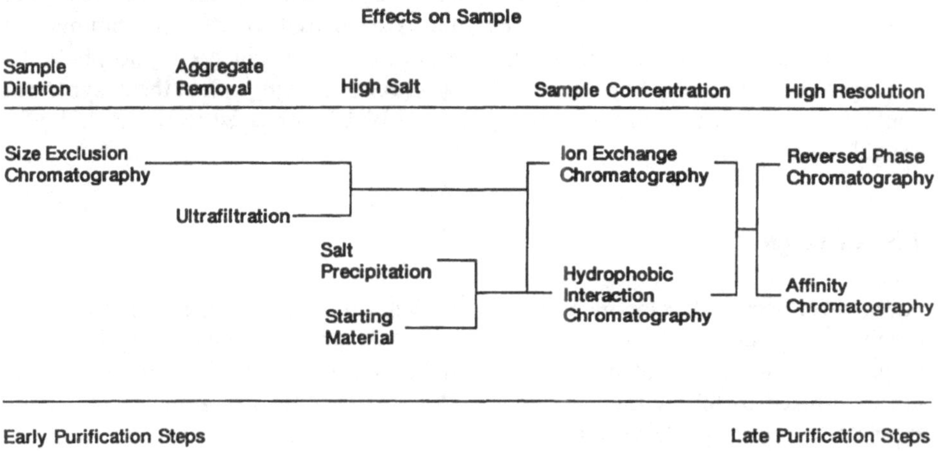

Fig. 1. The sequence of purification steps and their effects on sample conditions are illustrated as a guide for designing a purification strategy

the effects various separation techniques have on the protein sample. These effects should be considered with regards to the sequence of the different techniques when designing a protein purification strategy. Diluting procedures such as ultrafiltration and size exclusion chromatography are not only effective in removing aggregates that may interfere with subsequent chromatographic steps but also minimize the effects of proteases that may be present in the starting material. In some cases, therefore, these procedures may prove to be beneficial if used early in a purification scheme. High resolution techniques such as reversed phase and affinity chromatography can then be employed later to achieve final purity. Occasionally, however, concentrating methods (ion exchange, reversed phase, hydrophobic interaction and affinity chromatography) may be more appropriate at an initial stage particularly if the protein one wishes to purify is present at a low concentration and/or in a large volume of beginning material. Proteolysis can then often be substantially reduced by addition of appropriate inhibitors (see below).

Another factor to be considered when designing the order of steps is the condition in which the protein emerges from a particular step. Hydrophobic interaction chromatography may be a useful fractionation procedure early in a purification scheme when the sample is in a high ionic strength aqueous solution, for example following ammonium sulfate precipitation or isolation from a culture medium containing high levels of salts [36]. Reversed phase or affinity chromatography following hydrophobic interaction chromatography can then be used to both desalt and further purify the protein of interest [37]. However, desalting steps which do not result in further sample purification should be avoided since they are time-consuming and often lead to an overall reduction in final yield. Indeed, when performing sequential purification steps utilizing differential ionic strength buffers, it may be worthwhile to experiment with dilution of the eluting salt concentration or direct loading of a fraction from a previous step without desalting. Vardanis [38], for example, observed increased resolution by sequential ion exchange chromatography when the sample was reapplied in the eluting salt from the previous column as compared to when the sample was first desalted. Similarly, a 50% increase in purification of glucose-6-phosphate dehydrogenase was achieved by anion exchange chromatography when the enzyme was applied in an aqueous buffer containing 0.38 M NaCl compared to the same buffer devoid of salt [39].

2.3 Assessment of Purity

An integral part of developing a protein purification strategy is the availability of a monitoring system to follow the elution profile of the protein from a given chromatographic column. For proteins with known biological activity, specific assays are simply utilized to follow the purification. Proteins with no recognized activity, for example structural proteins, require an alternative strategy for their identification. The availability of an antibody, either monoclonal or polyclonal, to such proteins allows the use of a detection system based upon the antigenic properties of the protein. However, in this case one often finds oneself in a circular set of circumstances whereby an antiserum cannot be prepared without the purified protein and purification cannot be achieved without an antiserum. One recent advance that begins to break this circle is the use of antisera prepared against synthetic peptides whose amino acid sequence is determined from the nucleotide sequence of the corresponding complementary DNA (cDNA) or genomic DNA. However, once again this may be inappropriate since it is often the objective of the purification scheme to prepare a protein for amino acid sequence analysis from which synthetic oligonucleotide probes are generated to isolate the corresponding gene.

The assessment of protein homogeneity occupies a cornerstone in the purification strategy not only for its ability to analyze the purity of the final product but also for its role in developing a monitoring system. Traditionally, electrophoretic techniques have complemented chromatography as the method of choice for assessing purity. One-dimensional discontinuous polyacrylamide gel electrophoresis (PAGE) in the presence of sodium dodecyl sulfate (SDS) is a powerful method of resolving complex protein mixtures on the basis of their molecular weight. Isoelectric focusing (IEF) separates proteins on the basis of their net charge. Two-dimensional gel electrophoresis was developed to increase the resolving power of conventional electrophoresis by first separating proteins according to their net charge on IEF gels followed by SDS-PAGE. Once separated, proteins are visualized by Coomassie blue or silver staining. Combined two-dimensional SDS-PAGE followed by silver staining has the potential to separate and identify over 100 different proteins on a single gel [40]. Indeed, the high resolving power of two-dimensional gel electrophoresis makes it an ideal choice as a detection system for a protein purification strategy. Anderson and

Anderson [41] have advocated a rational approach to protein fractionation whereby the starting material is subjected to a variety of different small-scale analytical procedures from which the protein of interest is identified on two-dimensional gels (Fig. 2). The procedure that produces the greatest molecular weight and charge separation from contaminants can then be scaled up to a preparative mode followed by subsequent separations according to size and/or charge parameters theoretically yielding a pure product after a three- or four-step protocol. As technological developments continually reduce the amount of protein required for visualization, these electrophoretic methods will occupy a larger role in the future of protein purification.

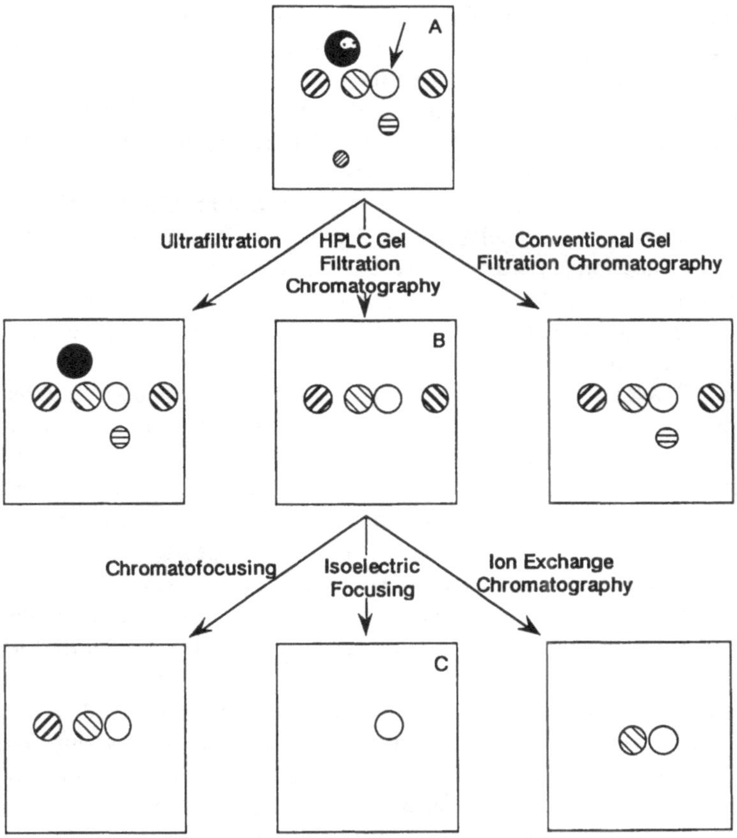

Fig. 2. Designing a rational approach to protein purification [41]. In this hypothetical example, two dimensional polyacrylamide gel electrophoresis of the starting material (*A*) reveals a large array of proteins in addition to the component of interest (indicated by the *arrow*). A variety of small-scale molecular weight separation techniques are tested and HPLC gel filtration chromatography is found to be most effective in removing contaminants (*B*). Several analytical charge separation techniques are then tested of which isoelectric focusing is found to produce purification to homogeneity (*C*). In this case, the final purification scheme would be increased to preparative scale HPLC gel filtration and isoelectric focusing

3 Choice of Starting Material

3.1 Relative Protein Abundance

Deciding which initial steps to use in a purification scheme depends upon both the concentration and physical state in which the protein exists in the starting material. Proteins initially present at relatively high concentrations in the starting material are normally easily purified. For example, Fox et al. [42] employed a four-step purification scheme consisting of DEAE-Sephadex chromatography, ammonium sulfate precipitation, phenyl-Sepharose and ion exchange chromatography to isolate an enzyme constituting 2.5% of the total soluble protein of *Methanobacterium thermoautotrophicum*. Proteins present at only trace amounts, on the contrary, are often extremely difficult to purify particularly if a contaminant is present in the starting mixture at a relatively high percentage of the total protein content. However, the same component of interest when present at a higher or equal abundance in a different source (e.g. different tissue) that contains a large variety of constituents but none at a high concentration may be easier to purify. Three examples can be cited to illustrate this point. First, preparative-scale isolation of the human interferon-γ receptor was achieved after identification of placenta as a crude starting material that expressed greater than 30 times the amount of receptor compared to cells grown in tissue culture [43]. Second, the envelope glycoprotein of human T-cell leukemia virus type I was purified to homogeneity using virus infected cell extracts rather than purified virions as starting material [44]. Although this virus contains less than 15 proteins compared to the thousands present in its host cells, it was found to be a poorer source for isolating the envelope protein in part because this protein dissociates from the virus during density gradient centrifugation and was therefore present at a relatively lower amount compared to infected cells. Finally, purification of the low molecular weight rabbit plasma protein α_1-microglobulin was aided by the use of nephrotoxic chemicals [45]. Renal tubular damage induced by sodium chromate caused a low molecular weight proteinuria characterized by the absence of larger plasma proteins (e.g. abumin and immunoglobulins) and relatively enriched for α_1-microglobulin thus facilitating its isolation.

In an analogous vein, the isolation and characterization of novel biologically active polypeptides and proteins has recently been achieved with the advent of serum-free tissue culture media. Two examples are noteworthy. First, serum-free media have provided assay systems used to monitor the purification of a variety of protein growth factors without interference from endogenous factors present in serum. MTW9/PL2 rat mammary tumor cells maintained in serum-free culture were used to isolate a growth promoting activity from porcine pituitary extracts which was subsequently identified as transferrin [46]. Lactoferrin was similarly purified and found to possess mitogenic properties for human lymphocytes [47]. These reports complement other examples of purification of new tissue culture derived growth factors, including those derived from epidermal [48], platelet [49] and fibroblast [50] cell culture. Secondly, other cytokines present at nanomolar to picomolar concentrations in tissue culture media were purified after the development of serum-free culture conditions. Examples of these include a variety of interleukins [51,52], colony stimulating factors

[53,54], tumor necrosis factor [55] and a leukemia cell-derived factor [56]. It is therefore apparent that any procedure to enrich for a particular protein in the starting material can greatly enhance the ability to achieve its final purification.

3.2 Physical State of the Protein

The physical state in which the protein of interest exists in the original material also influences the initial purification strategy. Soluble proteins naturally found in biological fluids such as plasma, urine and saliva or excreted by cells into tissue culture media can be directly processed by various separations protocols. In contrast, proteins that are present inside cells or tissues require a preliminary extraction step prior to fractionation. This normally involves physical disruption of the cell by homogenization, sonication or freeze-thawing followed by a slow-speed centrifugation to separate debris from the soluble protein present in the supernatant fluid. Subcellular fractionation can subsequently be achieved by high-speed differential centrifugation to separate the nuclear, mitochondrial, lysosomal, microsomal and soluble components (Fig. 3). Assaying these individual fractions for the presence of the protein of interest often identifies the particular subcellular fraction from which to begin purification although exceptions to this generalization have been reported [57]. Furthermore, membrane-associated proteins may tend to aggregate and exhibit reduced solubility in aqueous buffers. To overcome these problems, a plasma membrane fraction can subsequently be prepared by high-speed centrifugation from which bound proteins can be extracted by the use of detergents [58]. It should be noted, however, that the choice of detergent can often dramatically influence the final protein recovery as illustrated by the studies of Nandi et al. [59]. These authors tested a variety of ionic and nonionic detergents under different extraction conditions and observed that very low concentrations of SDS were most effective in purifying an adenosine triphosphatase from fundic mucosa. Alternatively, some integral membrane plasma proteins have been found to be released into solution by mild proteolysis with a nonspecific protease such as papain which cleaves the hydrophilic extracellular region of the molecule from the lipophilic membrane spanning segment [60]. Preservation of enzymatic activity of human erythrocyte acetylcholinesterase [61] as well as the crystallization and x-ray diffraction studies of human histocompatibility proteins [62] isolated in this manner have verified the utility of this procedure in producing active and homogenous protein fragments.

It is easier to purify a protein if a good source of it is available. The physiological state and growth conditions of cells affect their level of protein synthesis. It is, therefore, always worthwhile to determine the growth conditions where the rate of synthesis for the protein of interest is the highest and the presence of degradative enzymes (proteases, etc.) is minimal. Generally, the material should be obtained during the early stages of protein synthesis but one should consider that in this case the final yield may be reduced. For example, Fig. 4 illustrates a kinetic analysis of human immunodeficiency virus proteins synthesized following infection of human lymphocytes. In this case, a clear relationship exists between several precursor polyproteins and their normal proteolytic degradation products. The major external glycoprotein, gp160, is converted to a smaller polypeptide, gp120, in a time-dependent manner.

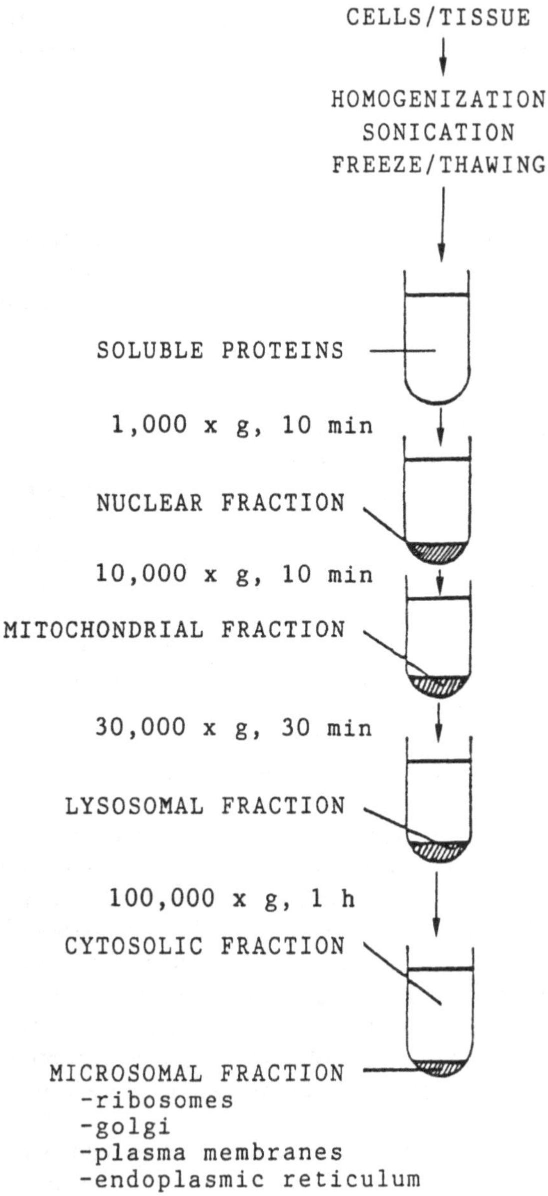

Fig. 3. Differential centrifugation of a cell or tissue extract can be used to fractionate cellular components. Following solubilization of cells and removal of debris by filtration, the soluble proteins can be differentially recovered by centrifugation at successively increasing centrifugal forces

Similarly, the polyprotein p55 is degraded to at least two products (indicated in Fig. 2 as p24) via an intermediate p41 moiety. The purification of gp120 or p24 would therefore be dependent upon the correct kinetic conditions of their respective syntheses.

Individual proteins are degraded at different rates depending on the physiological state of the cell. Damaged and abnormal proteins are quickly degraded in vivo as are many regulatory proteins. Regulation of the rates of synthesis and degradation of metabolically unstable proteins determines their cellular concentration. Adenosine

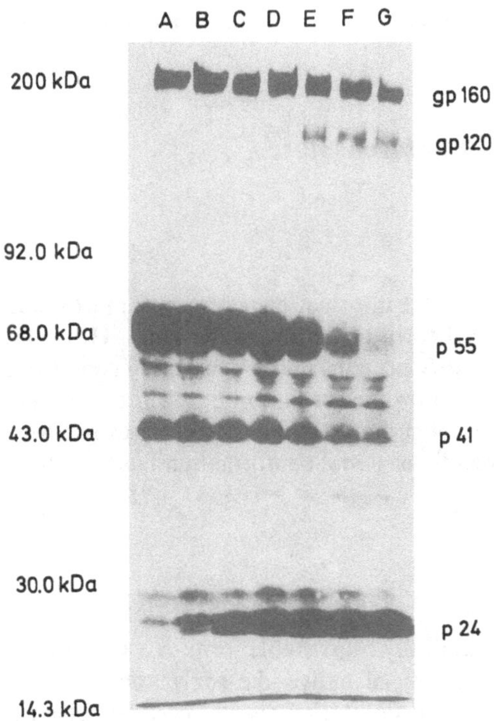

Fig. 4. Kinetic analysis of proteins synthesized by the human immunodeficiency virus type 1 (HIV-1). Human lymphocytes were infected in vitro with HIV-1, labeled for 20 min with [35S]methionine, resuspended in fresh medium, incubated for various lengths of time, lysed in detergent solution and HIV-1 proteins isolated by immunoprecipitation and analyzed by SDS-PAGE. Lanes *A* to *G* illustrate results of 10, 20, 40, 60, 120, 180 and 240 min incubations in fresh medium. The positions of molecular weight markers are indicated on the left in kDa. Viral proteins are indicated on the right

triphosphate dependent intracellular protein turnover involves covalent conjugation of ubiquitin to labile intracellular proteins. Furthermore, the recognition of an amino-terminal residue in a protein appears to mediate its in vivo half-life [63]. The in vivo half-life for 6-methylsalicylic acid synthetase, the first enzyme of the patulin pathway, in *Penicillium urticae* was much shorter (7 h) than those of two later enzymes, *m*-hydroxylbenzyl alcohol dehydrogenase and isoepoxydon dehydrogenase (17 and 19 h, respectively). The mechanism of inactivation of many such enzymes involved in secondary metabolism is unknown [64]. Incubation of gramicidin synthetase at 37 °C for 1 h resulted in loss of 85% of its activity. This inactivation in agitated cultures, harvested cells as well as cell-free extracts, was at least partly due to the presence of oxygen [65]. The gramicidin S synthetase complex is present in *Bacillus brevis* at the beginning of idiophase, reaches a peak and then diminishes while cells are producing gramicidin S. A nitrogen overlay to exclude oxygen stabilized the activity for a few hours. This thiol-mediated loss of activity was minimized by maintenance of aerobic conditions.

Several different proteases are present in *P. urticae* during patulin production [64]. 6-Methylsalicylic acid synthetase was stabilized by a combination of NADPH, acetyl coenzyme A (CoA), malonyl-CoA, phenylmethylsulfonyl fluoride (PMSF) and dithiothreitol. Substrates and protease inhibitors stabilized the activity of enzymes involved in penicillin biosynthesis in cell-free extracts [66,67]. The balance between the rates of synthesis and degradation of these enzymes in secondary metabolism affects their level in the cell. As a result, growth conditions with optimum physiological state of the cells must be standardized to obtain maximum enzyme synthesis.

4 Purification of Proteins in the Denatured State

4.1 Protein Native State

The amino acid sequence of a protein contains the information necessary for its folding into a biologically active three-dimensional form called the native state. However, in some cases it may be advantageous to purify proteins under denaturing conditions whereby conformation is abolished to maximize the resolution between the protein of interest and contaminants. Purification of proteins in denatured states followed by renaturation to a properly folded three-dimensional conformation has now been achieved for a number of proteins [5].

4.2 Denaturation-Renaturation

The native state is usually the thermodynamically most stable form of a protein although it usually interconverts among a variety of native-like enzymatically active specific three-dimensional globular conformations. Many agents (organic solvents, heat, detergents, chaotropic salts, pressure) destabilize the native state of a protein resulting in the protein's loss of biological activity due to a change in conformation. However, these biologically inactive proteins can be renatured to their native state by slow removal of the denaturant [68]. Slow dialysis in physiological buffer allows the protein to "shuffle" its way back to the native state [69]. Some protein folding steps are quite slow and require several days to reach completion. Rates of dilution and denaturant concentration may dictate the pathway of refolding [70-72]. The σ subunit of RNA polymerase, DNA topoisomerase [73], 1,25-dihydroxyvitamin D3 receptor [74], the α subunit of torpedo California acetylcholine receptor [75], estrogen receptor [76], chymotrypsin [77], urease [78], fructosyltransferase [79] and lactamase [80] have been renatured to biologically active native states after elution from SDS-PAGE gels [81,82].

Hydrophobic column packings can denature certain proteins [83,84]. Denaturation can be overcome by minimizing on-column dwell time, lowering the temperature and using mild denaturant solvents such as triethylammonium phosphate [85,86]. Improved recovery of hydrophobic proteins has been obtained by including denaturants in the mobile phase of reversed phase columns [87]. Use of urea, guanidine hydrochloride, SDS and organic solvents avoids skew elution patterns in size exclusion chromatography [88,89]. Viral coat proteins of Sendai virus [90], subunits of α crystallin [91] and tubulin [92] have been purified by using denaturing solvents in combination with high-performance ion exchange chromatography.

Detergent extraction and purification of membrane proteins has been reviewed [93-95]. Detergents can prevent proteolysis during crude extract preparation. However, many commerical preparations of detergents (e.g. Brij, Tween, Lubrol, Triton) are mixtures of related compounds of different molecular weight and contain peroxides and carbonyl components which react with protein sulfhydryl groups significantly changing their tertiary structure or activity during purification procedures [96]. Ionic and hydrophobic detergents can be removed from proteins by ion exchange chromatography and hydrophobic interaction chromatography respectively [97]. Dialysis, ultrafiltration, solvent extraction and gel filtration can be employed to remove detergents after dissociation from proteins by using acidic conditions [98,99]. A strongly bound detergent such as SDS can be competed off by an excess of a weaker detergent. Acetone, ethanol or acid have been used to precipitate proteins out of detergent solutions [73,100]. Alternatively, organic or acidic solvents can precipitate proteins directly out of solution without the use of detergents [101,102].

Many disulfide-containing proteins precipitate in reduced form. Attempts to overcome this problem have involved selective reduction of cystine residues or covalent modification at other amino acids prior to purification. Certain disulfides of human antihemophilic factor were selectively reduced before purification with thiol-disulfide interchange chromatography. Denatured myelin basic protein bound irreversibly to gel filtration media but proteolipid did not [103]. Covalent addition of polyethylene glycol and other hydrophobic agents allowed extraction of these types of proteins into organic solvents [104]. Addition of charged groups can also increase the water solubility of certain proteins.

The precipitation or chromatographic behavior of a protein cannot be always predicted by its amino acid sequence. After reduction of disulfide groups, bovine growth hormone is extremely insoluble in water while growth hormones of other species with high sequence homology are soluble. Local surface charges on the proteins which are independent of the net charge could affect the retention of proteins on ion exchangers near the isoelectric point [105]. Many proteins have asymmetric surface groups. For example, the DNA binding acidic proteins have a region of basic surface charge which is responsible for their binding to DNA and phosphocellulose [5]. Inclusion of denaturants eliminates the surface charge and allows accurate determination of molecular weights by size exclusion chromatography [88,89,106].

4.3 Inclusion Bodies

Many eukaryotic proteins accumulate in bacteria as fully reduced and denatured aggregates called inclusion bodies. The presence of inclusion bodies in genetically engineered bacterial cells and various purification methods to recover active proteins from them have been recently reviewed [106a]. In some cases the formation of inclusion bodies may actually simplify their purification. The majority of contaminants are washed out after lysing the cells followed by solubilization of inclusion bodies in a denaturation and renaturation buffer [107,108]. Figure 5 illustrates the extraction of a recombinant viral protein from inclusion bodies of *Escherichia coli*. In this case, urea was found to be the superior extracting agent to produce the protein in a state amenable for further purification and analysis. Detection of activity in situ in SDS-gels and renaturation of the protein may be enhanced by inclusion of a natural substrate as a template in the renaturation buffer [80,109]. Alternatively, renaturation procedures may be unnecessary depending on the goal of the purification scheme.

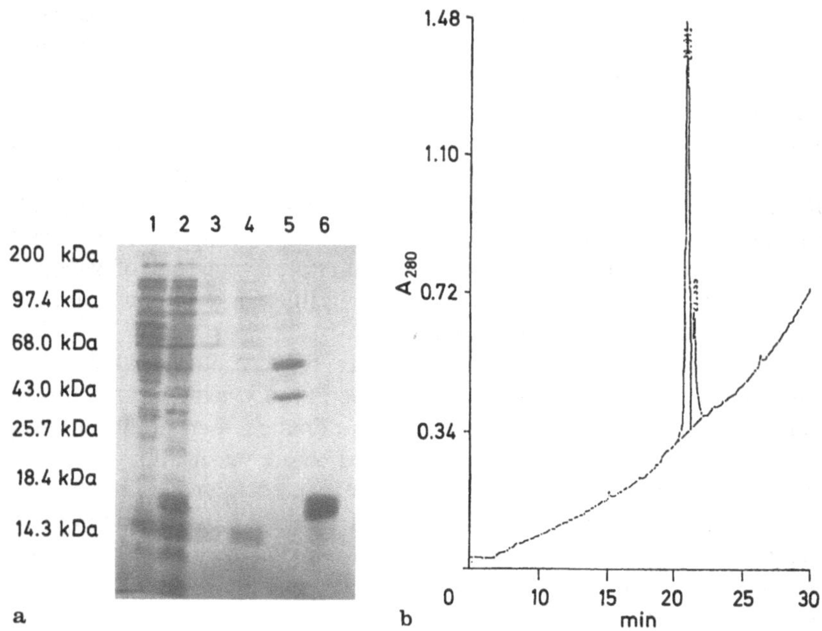

Fig. 5a, b. Purification of a recombinant protein from inclusion bodies. A segment of the genome of the human T-cell leukemia virus type I encoding the transmembrane envelope protein was cloned into plasmid pBR322 and used to transform *E. coli* strain HB101. Panel **a** illustrates a Coomassie blue stained SDS-PAGE analysis of a negative control extract of uninduced cells (*lane 1*), an extract of cells induced with heat (*lane 2*), the supernatant remaining after lysozyme extraction (*lane 3*), the inclusion body pellet following lysozyme treatment subsequently extracted with deoxycholate and Triton X-100 (*lane 4*), followed by octyl β-glucoside (*lane 5*), followed by 8 M urea (*lane 6*). The positions of molecular weight markers are indicated on the left in kDa. A major band at 16,000 Da was observed following urea extraction from which minor high molecular weight contaminants were removed by size exclusion chromatography on a Sephadex G-75 column. Panel **b** reveals that a single, prominent peak was observed when this preparation was analyzed by reversed phase HPLC indicative of its homogeneity. Courtesy of Dr. Chang-Chih Tai, Biotech Research Laboratories and Dr. Mark Connelly, DuPont

For example, since many recombinant proteins are produced in denatured form, antibodies raised against denatured protein may be more effective than antiserum against the native state for screening a cDNA library for expression of eukaryotic proteins [110].

Proteins can be renatured by the slow dialysis or dilution method after complete denaturation to random coils in a solution of chaotropic salt. Concentrated urea solutions, which yield proteinreactive cyanate ions and guanidine salts are, therefore, preferable. Proteins purified by SDS-PAGE can be eluted into SDS-containing buffer by diffusion, precipitated with acetone and denatured [73,111]. Inclusion of detergents in the refolding buffer or pretreatment of proteins with detergents may counteract aggregation and thus facilitate protein renaturation [72]. Renaturation of proteins containing disulfides has also been achieved [112-114].

5 Preparation of Cell-Free Extracts

For preparing cell-free extracts, the methods of choice depend upon the state of the material. Animal organs (liver, kidney, etc.) and various plant tissues can be homogenized in a "Waring blender" with a buffer after freezing. Bacterial cells are broken either by freezing and thawing, grinding with alumina, glasss beads or sand, sonication or use of a "French press". There are various types of mills commercially available to break microbial cells. The buffer should contain 20% glycerol or ethylene glycol to stabilize enzyme activity. Many succesful purification schemes have relied upon the presence of several different protease inhibitors particularly during steps which denature proteins. For example, the following mixture has been recommended as a starting point [115]: 1 μM leupeptin (for serine and thiol proteases), 200 μM phenylmethylsulfonyl fluoride (PMSF, for serine proteases), 1 μM pepstatin (for acid proteases) and 100 μM EDTA (for metalloproteases). Protease inhibitor solutions should be prepared fresh daily and added at each step of the purification scheme since many are unstable in aqueous solution (e.g. PMSF). Furthermore, many inhibitors are only effective against active enzymes and will not affect proteases that naturally exist as inactive complexes with endogenous inhibitors but which are activated upon dissociation. Polyvinyl polypyrrolidone and nonionic polymeric absorbents such as amberlite XAD-4 (Sigma Chemical Co., St. Louis, MO) are added to prevent inactivation of proteins by phenols.

6 Bulk Precipitation and Phase Extraction

Once the protein has been solubilized, an early purification step often involves salt or alcohol mediated precipitation followed by resolubilization and desalting. The objective of this procedure is two-fold. The first is to reduce the volume of the starting material to a level amenable to subsequent purification steps. Second, gross impurities such as carbohydrates, lipids and nucleic acids are removed. Although the degree of purification achieved by precipitating techniques is modest, the contaminants that are depleted often interfere with later procedures [35]. Fractionation of protein mixtures by precipitation with organic solvents such as ethanol and acetone, ammonium sulfate and polyethylene glycol is often used. Ammonium sulfate is the preferred precipitant because of its high solubility (up to 4 M), low cost, and its stabilizing effect on the activity of most enzymes.

Protein precipitation by various agents has been reviewed [116]. The charged polymer polyethyleneimine (PEI) has been used in protein purification. PEI forms ionic complexes with and thus precipitates macromolecules containing acidic domains such as nucleic acids and proteins. Salt concentration, pH and the concentration of the material to be precipitated affect the degree of precipitation. Instead of using protamine sulfate or streptomycin sulfate, PEI would be used to remove nucleic acids and nucleoproteins from crude extracts followed by ammonium sulfate precipitation and chromatography. Usually clear supernatant fractions are obtained after formation of charge neutralization complexes and creation of cross bridges between complexes by PEI. The pKa value of the imino groups of PEI is 10 to 11 and fractionation is done at slightly alkaline pH.

Aqueous two-phase systems based on the behavior of hydrophilic polymers in solution provide an extraction technique for purification of enzymes on large-scale. Cell debris and the desired enzyme are partitioned into opposite phases independent of temperature. The selectivity of partition is further enhanced by covalent attachment of an affinity ligand for the desired enzyme to a phase-forming polymer and performing the extraction under conditions which allow ligand-enzyme complex formation. Polyethylene glycol (PEG)-salt systems which require high salt concentrations will dissociate affinity ligand-enzyme complexes. Affinity extraction at low ionic strength using PEG-dextran systems was applied for extraction of formate dehydrogenase from *Candida boidinii* [117]. A triazine dye was bound to PEG and the enzyme-ligand complex was extracted in the top phase. The complex was dissociated by the addition of salt expelling the ligand to the top phase and the enzyme to the bottom phase. Three-phase partitioning methods using ammonium sulfate and *t*-butanol for separating enzymes in mid-layer have been reported [118]. Besides purification of the desired enzyme, three-phase partitioning extracts apolar pigments into the *t*-butanol layer yielding a clean extract.

7 Conventional vs High-Performance Liquid Chromatography

7.1 The Principle of Liquid Chromatography

Chromatography is based upon the principle of differential partition of a specific component between two phases, the mobile phase and the stationary phase. The former serves to move the sample through the latter in a manner that optimizes the physical and chemical interactions of the sample with both phases. The mobile phase can exist in either the liquid or gaseous state while the stationary phase commonly occurs as a liquid or solid. Protein separations are normally performed in a combination refered to as liquid chromatography which employs a liquid mobile phase and a solid stationary phase packed into a tubular shaped column.

7.2 High-Performance Liquid Chromatography

The past ten years have witnessed a revolution in the development of new technologies directed towards the purification of proteins from complex biological fluids. Particularly important among these are high-performance liquid chromatographic methodologies. Compared to conventional liquid chromatography, HPLC offers the distinct advantages of superior resolution and speed of operation. This principle is portrayed in Fig. 6. Although traditional liquid chromatographic techniques are easy to use and operate at low financial cost, they are conducted on columns containing chromatographic media composed of carbohydrate or synthetic polymer supports with particle sizes generally in the range of 50 to 150 μm. As a result of these relatively large and heterogeneous sizes, these media provide for only poor resolution of protein mixtures. Furthermore, their susceptibility to volume compression due to high linear flow velocities necessitates slow flow rates and correspondingly long

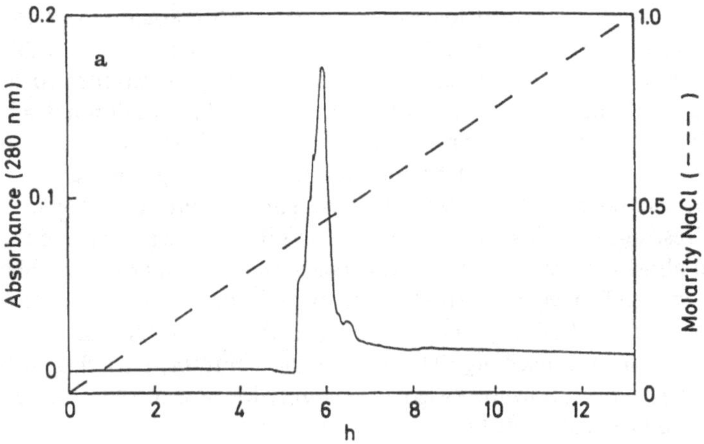

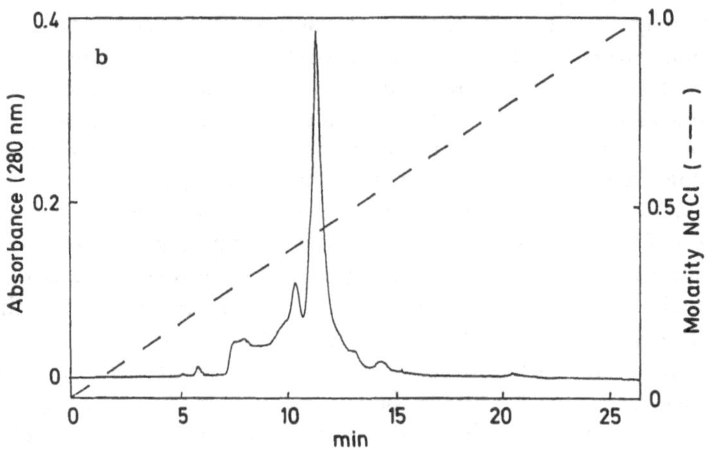

Fig. 6a, b. Comparison of conventional ion exchange chromatography with HPLC. A commercial preparation of ovalbumin was separated using either a standard, low pressure DEAE-cellulose column (1.5 × 15 cm; panel **a**) at a flow rate of 25 ml h^{-1} or a DEAE 5PW HPLC column (Waters; 2.15 × 15 cm; panel **b**) at a flow rate of 4 ml min^{-1}. A linear gradient from 0 to 1 M NaCl in 10 mM sodium phosphate buffer, pH 7.2 was used for elution. Note the increased peak resolution and reduced retention time produced with the HPLC column

separation times. Normally this requires that such separations be performed at reduced temperature in order to preserve the protein of interest from degradative physical, chemical and/or enzymatic effects.

HPLC, introduced in the late 1970's, employs silica- or polymer-based supports to which are attached various functional groups to overcome the limitations of conventional liquid chromatography [119–121]. These supports are available in a wide array of particle sizes that can be manufactured with a high degree of uniformity. As a result, the resolution of protein mixtures by HPLC far surpasses that even remotely possible by traditional chromatography. Silica-based supports also possess the capability to withstand high linear flow velocities without significant compression effects allowing separations to be performed on a time-scale of minutes rather than hours. As illustrated in Fig. 6, a typical ion exchange chromatographic separation by HPLC can utilize flow rates that are 10 to 15 times greater than those possible by conventional chromatography. Such separations may usually be performed at ambient temperatures without substantial degradative losses. Two additional advantages are provided by silica-based supports used in HPLC. First, the ability to modify the silica surface with a variety of chemistries provides the protein chemist with a large array of separation modes. Second, these particles are available in a wide range of pore sizes. Those with small diameter pores (100 Å or less) provide for large surface areas to allow increased interactions between the stationary and mobile phases. These are often used to separate small proteins and peptides by chromatographic techniques that utilize bonding interactions with the stationary phase, for example reversed phase and ion exchange HPLC. On the contrary, particles with large pore sizes (300 to 1,000 Å) allow increased accessibility for large proteins which makes them ideal for size exclusion HPLC.

The development of column packings with high performance, high resolution, speed, stability and rigidity has made it easier to isolate and purify proteins. Packings based on Sepharose and Toyopearl polymeric resins provide high flow rates and good recovery of proteins while preserving biological activity. Furthermore, like the silica-based resins employed in HPLC, their versatility allows them to accomplish size exclusion, ion exchange, chromatofocusing, reversed phase, hydrophobic interaction and affinity chromatography. Fast protein liquid chromatography (FPLC), developed by Pharmacia LKB Biotechnology, is such a chromatographic system. Utilizing matrices based upon Sepharose resins, it generally employs operating pressures between those of conventional and HPLC chromatographies. FPLC is therefore capable of combining the benefits of high resolution and speed characteristic of HPLC with biocompatibility, long column life and ease of scale up. Moreover, the modular design of the FPLC system allows the protein biochemist to assemble the specific configuration necessary for his or her needs. Since it is a total chromatographic system, all components from pumps to detector, recorder and fraction collector are designed into a single unit. Consequently, FPLC has gained world-wide popularity as for the purification of macromolecules, particularly peptides and proteins.

7.3 Mathematical Descriptions

A variety of mathematical formulas have been developed to describe the resolving characteristics of a particular chromatography system and these have been reviewed [122]. The most popular are those describing the concept of theoretical plates (N), which is basically a measure of the band or peak width around the center of a Gaussian elution peak. Thus,

$$N = (V/\sigma)^2 \tag{1}$$

where V is retention volume, time or distance and σ is a measure of peak variance. Sigma (σ) is correctly described algebraically in terms of classical statistics. More commonly, however, peak width is determined by any of a number of simpler techniques such as extrapolation of inflection point tangents to baseline or measurement at 4.4% of peak height. In either case, the larger the number of theoretical plates, the narrower the peak spreading that occurs. Column efficiency refers to the lack of band spreading as measured by the absolute number of theoretical plates for a given system or the height equivalent to a theoretical plate (HETP). If L is column length then

$$\text{HETP} = \text{L/N} \tag{2}$$

Resolution (R) is a measure of the separation of one component from another and is defined mathematically as the difference in elution volumes for two peaks divided by their average widths. The capacity factor (k′) measures the ratio of the total amount of component on the stationary phase to the total amount in the mobile phase. The distribution coefficient (K) refers to the same ratio using units of concentration. Both terms reveal where components elute relative to the void volume (V_0), the elution volume of an unretained substance, and are unaffected by flow rate or column dimensions. The selectivity factor (a) is equal to the ratio of the k′ values of two components and is a measure of how efficiently the mobile and stationary phases compete for a sample. The practical value of these terms lies in their ability to compare different chromatographic separations. For example, traditional low-pressure columns commonly possess HETP values of 100 to 300 theoretical plates per meter whereas a modern reversed phase HPLC column may exceed 100,000 plates per meter.

8 Descriptions of Chromatographic Techniques

In view of the dominance that HPLC technology has had over the field of macromolecular purification during the past decade, this section will be primarily devoted to recent developments in this area. A variety of HPLC methodologies exist that can

Table 1. Techniques of column chromatography

Separation mode	Characteristics
Size exclusion chromatography	Size separation and desalting of proteins, oligosaccharides and nucleic acids. Fractionation range from 100 to 50,000 Da.
Ion exchange chromatography	Purification of proteins and enzymes by charge. High resolution and speed with very high capacity. Compatible with preserving bioactivity.
Hydroxylapatite chromatography	High capacity with low degree of nonspecific binding. Unique separation parameter not primarily dependent on size, charge or hydrophobicity allowing resolution of complex mixtures when other methods fail.
Hydrophobic interaction chromatography	Purification of proteins by hydrophobic nature. Best suited when sample is at high ionic strength (e.g. after salt precipitation).
Reversed phase chromatography	Purification of peptides, fatty acids and small molecules by liquid-liquid partition.
Affinity chromatography	Very specific purification of proteins utilizing their biological affinities to ligands bound on solid support.

be applied to a given scheme to purify an individual protein from a complex mixture of components. Different techniques are based upon different physical or biological properties exhibited by polypeptides and proteins such as their hydrodynamic volume, ionic character, hydrophobicity (surface and interior) and ligand affinity. The following section will highlight the theoretical and practical considerations relating to these methods, viz. size exclusion, ion exchange, hydrophobic interaction, reversed phase and affinity chromatography. Table 1 lists the types of column chromatography techniques. Table 2 presents the commercial suppliers of conventional chromatography resins and HPLC columns.

Table 2. Commercial suppliers of chromatography resins and columns

Supplier	Conventional resins					HPLC columns					
	SEC	IEC	HAC	HIC	AC	SEC	IEC	HAC	RPC	HIC	AC
Baker		×		×	×		×		×		×
Beckman						×	×		×	×	×
BioRad	×	×	×		×	×	×	×	×	×	×
Brownlee (ABI)							×		×	×	
Calbiochem			×		×						
DuPont						×	×		×		×
EM Science	×					×	×		×	×	
Hamilton							×		×		
IBF Biotechnics	×	×			×						
ISCO						×	×		×	×	
Pharmacia LKB	×	×		×	×	×	×		×		
Phenomenex		×			×	×	×		×	×	
Pierce	×	×			×		×		×		×
Ranin						×	×	×	×	×	×
Separations Group							×		×		
Sigma	×	×	×		×				×		
Supelco						×	×	×	×	×	×
Synchrom						×	×		×	×	
TosoHaas	×	×		×	×	×	×	×	×	×	×
Waters (Millipore)						×	×		×	×	
Whatman		×					×		×		

SEC: Size exclusion chromatography;
IEC: Ion exchange chromatography;
HAC: Hydroxylapatite chromatography;
RPC: Reversed phase chromatography;
HIC: Hydrophobic interaction chromatography;
AC: Affinity chromatography

8.1 Size Exclusion Chromatography

Historically, size exclusion (gel filtration, gel permeation) chromatography was the first of the chromatographic methods to be adapted to HPLC [119,123]. Macromolecules in the mobile phase differentially penetrate the pores of the stationary phase particles according to their hydrodynamic volume such that the larger components are excluded from the interior of the particles whereas the smaller molecules are accessible to this volume. Thus, under ideal conditions, the order of elution can generally be predicted by the size of the protein since a linear relationship exists between its elution volume and the logarithmic value of its molecular weight. Operationally, however, some degree of interaction between the protein and size exclusion support cannot be avoided although these effects can be minimized by addition of modifiers such as salts [124] or detergents [125] to the mobile phase. A recent study on the chromatographic behavior of polypeptides and proteins in the range between 200 and 200,000 Da by size exclusion chromatography in aqueous solution concluded that the effects which determine their relative separations are dependent upon their molecular weight [124]. Thus, proteins with molecular weights greater than 3,000 Da were resolved predominantly based upon their hydrodynamic radii. In contrast, for polypeptides less than 900 Da separations were mainly influenced by their ionic character and hydrophobicity. Bennett et al. [126] observed that these ionic and polar interactions could be minimized by utilizing a mobile phase consisting of 40% aqueous acetonitrile containing 0.1% trifluoroacetic acid (TFA). This system allowed accurate estimation of molecular weights of small proteins in the range of 500 to 15,000 Da which was attributed to the hydrophobic properties of acetonitrile and the ion-pairing properties of TFA inhibiting nonpolar and polar interactions with the support. Although the application of high-performance technology to size exclusion chromatography has greatly enhanced its resolving power, this technique still remains one of the least effective means of separating protein mixtures in the armamentarium of HPLC. Proteins with a difference in molecular weight of 10% or less often cannot be separated. By comparison, peptides with a single amino acid substitution can be resolved by reversed phase HPLC [14].

A common practice used to increase the theoretical plate number (N) of size exclusion separations is to connect in series two or more columns containing the same pore size support. Similarly, since the effective molecular weight fractionation range is determined by the pore size rather than the column length, connection of two columns with differing pore size supports in tandem will increase not only N but also the fractionation range (Fig. 7). Most commercially available columns will separate proteins over about 1.5 to 2.5 orders of magnitude in size. However, by connection of two columns containing supports with pore sizes differing by a factor of 10, the effective separation range can be doubled [127]. Furthermore, this method of bimodal coupling produces more linear calibration curves than using a single column containing a mixture of particles with different pore sizes. Operationally, the columns should be connected in order of decreasing pore size and the individual calibration curves of each column should minimally overlap to achieve the maximum combined fractionation range.

Size exclusion chromatography is an excellent technique to employ early in a purification scheme, ideally as the first or second step, since it results in sample dilution. Proteolysis due to chemical or enzymatic effects can therefore be minimized at an early stage in purification with a resulting higher yield in the final product. A second clear advantage of size exclusion chromatography is the ability to select from a wide variety of mobile phases. Generally, mobile phases are buffered, low salt aqueous solutions (e.g., 0.25 M sodium acetate containing 0.25 M sodium chloride, pH 5.5) used to maintain native structure. However, extreme conditions of pH as

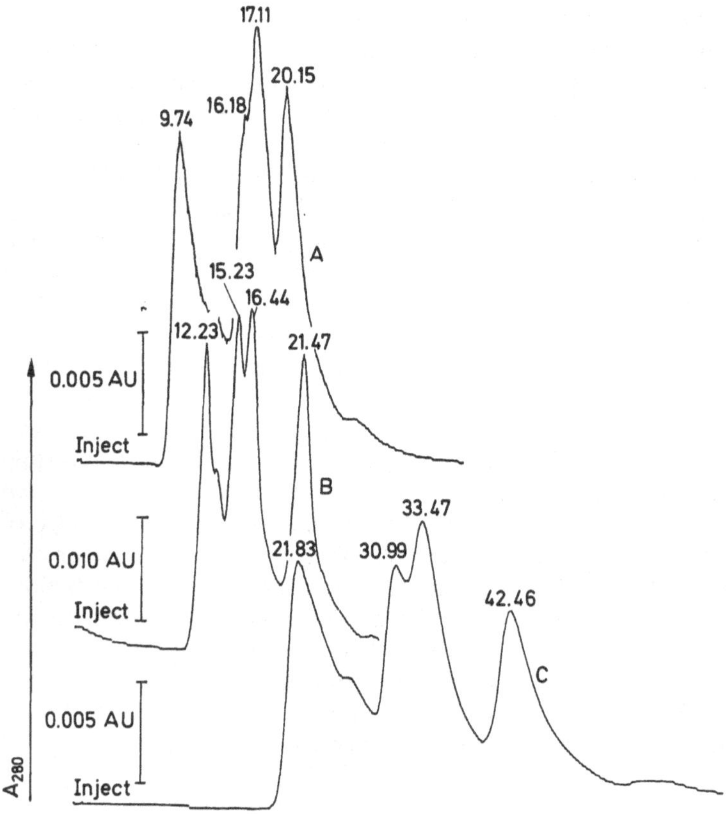

Fig. 7. Connection of two HPLC size exclusion columns in series. A standard protein mixture consisting of blue dextran (2,000,000 Da), bovine serum albumin (68,000 Da), ovalbumin (45,000 Da) and chymotrypsinogen (25,000 Da) was resolved on a Protein Pak 300 SW column (*A*; 0.75 × 30 cm; fractionation range 10,000–400,000 Da; Waters), a Protein Pak 125 column (*B*; 0.75 × 30 cm; fractionation range 2,000–80,000 Da; Waters) or the two columns connected in series (*C*). The flow rate was 0.5 ml min^{-1} using 10 mM Tris, pH 7.5 containing 0.5 M NaCl. Retention times in min of each protein are given above the peaks

well as the presence of organic solvents, high concentrations of salts and/or denaturing agents such as urea or guanidine may be employed to minimize aggregation effects while maintaining sample solubility [7,128,129]. Under these conditions, protein activity may not be recovered. Recent widespread use of both ionic and nonionic detergents has, in many cases, allowed proteins with unusual solubility properties to be isolated. Particularly illustrative of this is the purification of integral plasma membrane proteins [130–132]. Ionic detergents (e.g. SDS) are not as well suited as nonionic ones for extraction of membrane proteins because they are known to denature proteins. Other notable examples of the utility of size exclusion chromatography over other HPLC techniques are the study of protein folding and stability [133] and the dissociation constants of oligomeric polypeptides [134].

8.2 Ion Exchange Chromatography

Ion exchange chromatography of proteins involves the interaction of charged functional groups in the sample with ionic functional groups of opposite charge on the adsorbent surface. Two types of electrostatic interactions can be distinguished. Anion exchange chromatography is mediated by the interaction of negatively charged amino acid side chains (e.g., aspartic acid and glutamic acid) with positively charged supports. Proteins with a net negative charge, i.e. in an aqueous solution with a pH above their isoelectric point (pI), will bind to these supports. Cation exchangers are prepared by polymerization of negatively charged groups to silica for interaction with positively charged residues (e.g., lysine and arginine). Proteins in a solution with a pH below their pI will interact with these adsorbents. Further differentiation among ion exchange resins can be made according to their ionic capacity, commonly referred to as "weak" and "strong" although these terms do not refer to the strength of binding between the sample and matrix. Thus, a weak anion exchange resin would be provided by the diethylaminoethyl (DEAE) group whereas the quaternary aminoethyl (QAE) group is a strong anion exchanger. Examples of weak and strong cation exchangers are the carboxymethyl (CM) and sulfoxypropyl (SP) groups, respectively. Phosphocellulose cation exchange chromatography utilizes the phosphoryl group which is of intermediate strength compared to CM and SP modified supports. Cellulose phosphate resins have also been prepared to provide bifunctional cation exchangers containing both fully ionized and weak cation exchangers containing both fully ionized and weak cation groups. The advantage offered by this adsorbent is that it offers three different mechanisms of protein binding and elution [134a)]. First, it may be used as a normal cation exchange resin. Second, it may be used as an adsorbent to purify proteins with a specific affinity for phosphate groups [135)]. Finally, the substrate elution technique allows one to bind an enzyme to the resin and elute it with a protein substrate or inhibitor [136)].

The binding energy between the sample and adsorbent is dependent upon the solvent pH, ionic strength and selectivity of the counter-ion. Thus, bound components can be differentially eluted by alterations in any of these three parameters. Common practice has relied on constant conditions of pH and counter-ions with a gradient of increasing salt concentration. These conditions are normally compatible with maintaining native biological or enzymatic activity. However, if such activity is adversely affected by extremes in ionic strength, gradients of pH or selection of a different salt counter-ion may be employed [137–141)]. Utilization of a different salt counter-ion also affects the differential resolution of components in a complex mixture. Thus, protein separation on an ion exchange HPLC column can be facilitated by either optimizing the buffer conditions for a given chromatographic run or by rechromatography of the protein of interest on the same column in the presence of an alternative salt counter-ion. Proteins that co-eluate under the conditions of one buffer system may be easily resolved by another system. For example, whereas reduced and oxidized forms of cytochrome C could not be separated on a weak cation exchange column using gradients of NaCl or $(NH_4)_2SO_4$, the same column produced a discernable resolution using NaOAc and baseline separation with KH_2PO_4 as the counter-ion [141a)].

Generally, ion exchange HPLC produces poorer resolution than reversed phase HPLC although it is superior in recovering biological activity. However, exceptions have been noted, as in the case where ion exchange chromatography separated three polypeptides of diptheria toxin A that differed only at their carboxyl termini and could not be resolved by reversed phase HPLC [37)]. Resolution of proteins that differ by only a single charge has also been reported [142)]. Many other examples of protein separations by HPLC on ion exchange resins can be cited. These include enzymes [143,144)], peptide hormones [145)], hormone receptors [146)], serum proteins [147,148)] and structural proteins [149,150)]. These citations should not be considered as inclusive but rather as a general guide to illustrate the widespread application of ion exchange chromatography to protein isolations. Indeed, a recent compilation of proteins separated by ion exchange FPLC lists 127 different polypeptides from over 175 literature references [151)]. Prominent among this list are mouse monoclonal antibodies. Figure 8 illustrates the large scale isolation of an IgG1 monoclonal antibody by anion exchange FPLC. This example is particularly noteworthy in light of the fact that this isotype cannot be readily purified by conventional protein A immunoaffinity chromatography (see below).

Chromatofocusing is a relatively new technique that, like ion exchange chromatography, separates macromolecules on the basis of their net ionic charge [152)]. However,

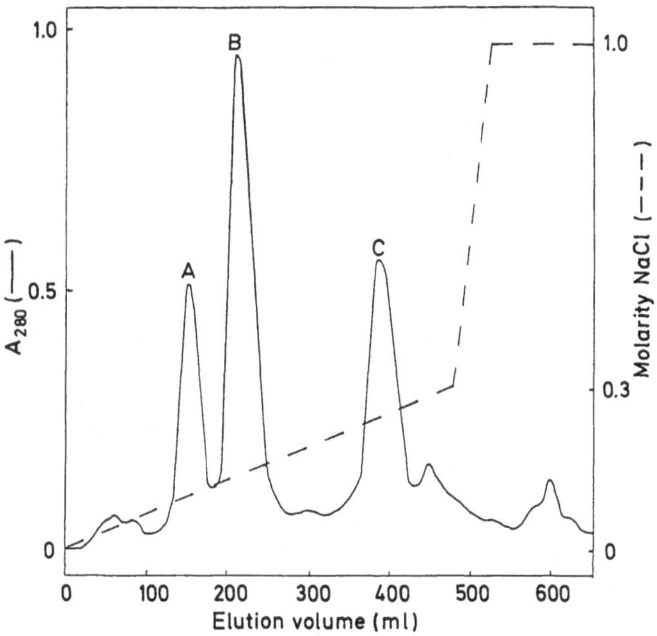

Fig. 8. Scale-up purification of a mouse monoclonal IgG1 antibody by ion exchange FPLC. Five milliliters of mouse ascites was loaded onto a Mono Q anion exchange column (0.5 × 5 cm; Pharmacia LKB Biotechnology) and the proteins eluted in 50 mM Tris, pH 8.6 with a 0 to 1.0 M gradient of NaCl as indicated at 10 ml min^{-1}. The major peak (*B*) eluting between 200 and 250 ml was observed by SDS-PAGE to contain >95% pure monoclonal antibody. Peaks *A* and *C* were identified as transferrin and albumin, respectively. Note: the isotype of this antibody precluded its isolation by protein A affinity chromatography. Courtesy of Dr. Steve Alexander, Biotech Research Laboratories

chromatofocusing, as its name implies, also relies on a focusing effect that is developed by cycles of sample adsorption to and desorption from the solid support. This principle is illustrated in Fig. 9. The sample is applied to the resin at high pH following with a buffer of lower pH containing various amphoteric components with a range of pKa values is used to titrate both the adsorbent and bound sample. The former

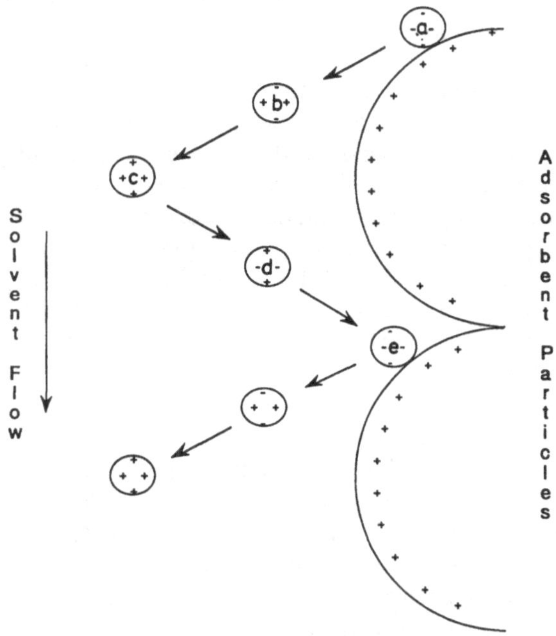

Fig. 9. The principle of chromatofocusing. The protein component with a net negative charge (*a*) is initially retained on the positively charged adsorbent. As the pH gradient is formed, the proteins at the rear of the sample zone are titrated first to neutrality (*b*) and then to a net positive charge (*c*) resulting in their repulsion from the adsorbent and subsequent migration into the front of the sample zone. Here the increase in pH results in their titration back to neutrality (*d*) and back to a net negative charge (*e*) leading to readsorption to the matrix and their return to the rear of the sample zone, completing the chromatofocusing cycle

results in a linear, descending pH gradient. As this pH gradient develops, the proteins at the rear of the sample zone are the first to be titrated to the same charge as the support and consequently desorbed from the support allowing them to migrate to the front of the sample zone. At this point they are titrated back to a charge opposite that of the adsorbent and readsorb to it. Consecutive cycles of exchange between the front and rear of the sample zone results in a focusing of this zone. Proteins comprising a crude mixture will effectively be separated based upon their isoelectric point (pI) with those possessing pIs near the initial column pH eluting first and those with pIs different from the initial pH eluting later. Like ion exchange HPLC, a multitude of different proteins have been isolated by chromatofocusing [153–156].

8.3 Hydroxylapatite Chromatography

Hydroxylapatite is a crystallized derivative of calcium phosphate originally developed by Tiselius and coworkers [157] for fractionation of complex protein mixtures. Proteins adsorb to it via ionic interactions and can subsequently be quantitatively eluted in a high ionic strength buffer. Since phosphate buffers are generally compatible with protein structure and function, increasing gradients of phosphate concentrations (e.g. 0.01 to 0.50 M) are commonly used to desorb bound components. It has been suggested that the actual mechanism of interaction between proteins and hydroxylapatite involves carbonyl and phosphate groups on the surface of native proteins [158]. Urea-induced protein denaturation was observed to decrease the binding to hydroxylapatite presumably as a consequence of a reduction in the number of these ionic groups on the surface of proteins in a random coil configuration [159]. However, others have observed that SDS, which also produces random coil denaturation, can be used in conjunction with hydroxylapatite chromatography for the separation of polypeptides [160]. Indeed, an advantageous feature of this method is that protein elution is independent of molecular weight thus making it a powerful companion to use following separation by size exclusion chromatography or polyacrylamide gel electrophoresis in the presence of SDS.

Although hydroxylapatite chromatography is based upon ionic adsorption, it provides a unique fractionation method with properties different from those of ion exchange chromatography. This fact, together with its high binding capacity, makes it a popular method of protein purification. Brooks [161] has reviewed the applications of hydroxylapatite chromatography to protein isolations including those of enzymes, chromatin proteins, glycoproteins, lipoproteins, phosphoproteins and lectins. Although this list is impressive, many practical problems have been encountered in handling this material, particularly those related to the inherent fragility of the crystals. However, careful attention to particle size allows high chromatographic resolution of proteins on this medium. Recently, hydroxylapatite HPLC columns have become commercially available and proven useful for the purification of peptides and proteins [162].

8.4 Reversed Phase Chromatography

In normal phase chromatography, the solvent is less polar than the adsorbent. Reversed phase chromatography uses a solvent that is more polar than the adsorbent. Silica is by nature very polar allowing normal phase chromatography to be performed directly on these particles. More commonly, however, silica is bonded with a polar functional group such as the cyano or amino groups and normal phase chromatography performed with a mobile phase such as hexane, methylene chloride, chloroform or ether. Silica, bonded with nonpolar functional groups such as alkyl or aromatic phenyl groups, is employed in reversed phase chromatography. In this case, the nonpolar sample consituents bind to the support pre-equilibrated in a polar solvent such as water or an aqueous buffer and are eluted with a mobile phase containing acetonitrile, methanol, isopropanol or tetrahydrofuran. Such elutions are performed utilizing either isocratic conditions (i.e., constant mobile phase composition) or gradients of increasing organic solvent. Stationary phases containing bonded alkyl groups (butyl, C-4; octyl, C-8; octadecyl, C-18) are the most popular reversed phase supports. Although there is minimal variation in the retention times of proteins on these different supports [163], better recoveries have been observed using the shorter carbon chain groups [164].

Reversed phase chromatography was initially applied to the separation and purification of small peptides since it was shown that even minor differences in their amino acid composition greatly affected their elution profile [165]. Furthermore, the relative content of hydrophobic amino acids can sometimes be used to predict their elution order. Conditions were developed that produced baseline separations of complex peptide mixtures, often by incorporating an ion-pairing agent (commonly dilute TFA) to elute components as one member of the ion-pair [166]. Reversed phase chromatography does, however, possess some handicaps particularly when attempting to isolate larger proteins. Prediction of elution order is often limited by secondary and tertiary structures that may prevent some hydrophobic amino acids from interacting with the support. Second, reproducibility of separations is often poor when comparing different commercial column sources or even comparing the same column from a single source [167]. Finally, native protein structure may be lost using standard organic solvents and acidic conditions of the mobile phase and hydrocarbon bonding in the stationary phase thus destroying biological activity or antigenic characteristics necessary for enzyme-based or immunological-based detection assays [83, 168–170]. The denaturative effects can be investigated on a small scale prior to applying a valuable sample to a reversed phase column by first preincubating it under conditions of the reversed phase chromatographic run. However, although denaturation may occur, post-column conditions of lyophilization or dialysis, in many cases, allow some degree of renaturation to occur. In fact, the list of proteins observed to undergo reversible denaturation following reversed phase HPLC is impressive. Selected examples of these include those of viral [171], bacterial [172], plant [173] and animal [145] origin. Figure 10 illustrates the recovery of biologically active cytokines isolated on a reversed phase column. In some cases, the native and denatured forms of a protein can be resolved allowing the kinetics of denaturation to be investigated as demonstrated by a beautiful series of papers by Karger and coworkers [83, 167, 172, 174].

The major advantages that reversed phase chromatography has over other methods of HPLC are its high degree of resolving power and its versatility. Both properties lead to a generalized scheme often employed in HPLC methods development [175]. First, one identifies the particular chromatographic elution parameters of the protein of interest, for example by determining its retention time during a gradient of organic solvent. Once this is established, a variety of factors can be changed to produce a baseline separation of the protein such that it is free of contaminants that elute nearby. Changing columns from different vendors may dramatically alter retention times in spite of the fact that particle size, pore size and carbon coverings may be identical. Alternatively, solvent optimization using mixtures of acetonitrile/propanol or acetonitrile/methanol [176] and chemical optimization by varying solvent pH [177] or ion-pairing agents are possible. For example, Fett and coworkers [178] observed that the purification of human angiogenin by reversed phase HPLC was dramatically improved by changing from a linear gradient of acetonitrile to a gradient of 2-pro-

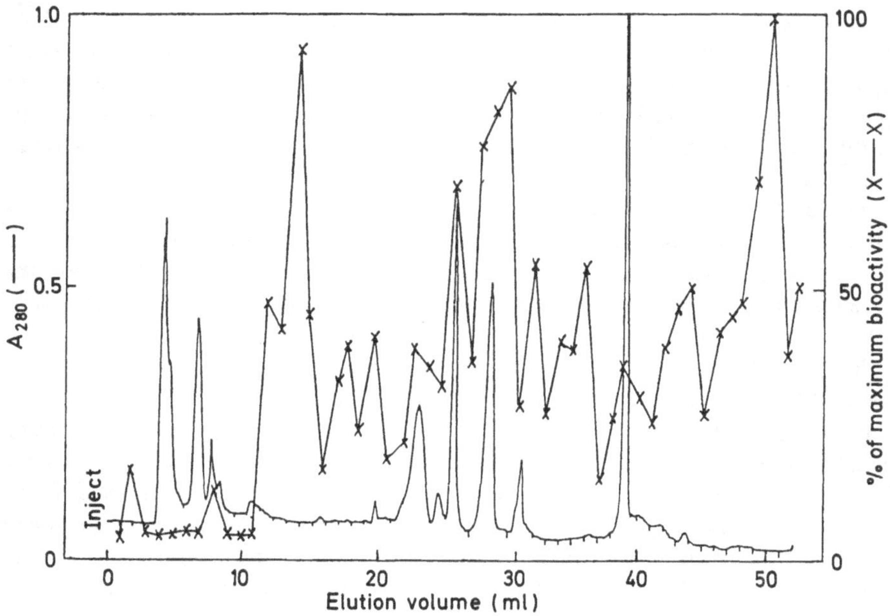

Fig. 10. Purification of biologically active cytokines by reversed phase HPLC. Conditioned medium from a Marek's disease virus transformed chicken T lymphocytic cell line was concentrated by lyophilization and resolved on an octadecyl (C-18) HPLC column developed with a linear 0 to 50% gradient of aqueous acetonitrile at pH 3.0. Individual fractions were concentrated by lyophilization, reconstituted in physiological buffer solution and assayed for bioactivity

panol/acetonitrile. Although this scheme is theoretically attractive, it is limited in its applicability by financial and time constraints. Testing a variety of different reversed phase HPLC columns from various suppliers can quickly develop into a very expensive undertaking with no guarantee of success. Furthermore, complete determination of the effects of solvent composition on resolution can be a time-consuming operation. In light of these restrictions, Berkowitz et al. [179] have proposed an integrated approach utilizing both conventional low pressure and high-performance liquid chromatographic techniques for protein purification. The first step of this scheme uses short (e.g. 5 cm) HPLC columns containing 3 to 5 μm silica-based packings to optimize a particular purification on an analytical scale. Low financial cost, quick analysis time and reduced sample consumption are the advantages procured when using these shorter columns compared to the standard 4.6 × 250 mm HPLC columns containing 5 μm particles. Once separation conditions have been optimized, inexpensive 40 μm silica-based particles are employed in the second step on a preparative-scale using traditional, low pressure chromatography hardware. These media permit high linear flow velocities without compressibility problems inherent in softer, carbohydrate-based packings and are thus ideally suited for use during the initial stages of protein isolations. Finally, 3 to 15 μm particles in standard-size HPLC columns are utilized in the third step. These provide high resolution separations required near the end of a purification scheme to remove contaminants that have copurified with the protein of interest due to similarities in their physical and chemical properties.

8.5 Hydrophobic Interaction Chromatography

The tendency of organic solvents to denature proteins during reversed phase HPLC separations has been complemented by the development of hydrophobic interaction chromatography. This has been a particularly important development in the field of protein purification since this denaturative effect requires that protein characterization by reversed phase chromatography alone be dependent upon some knowledge of their molecular weights [180]. As its name implies, hydrophobic interaction chromatography, like reversed phase chromatography, relies upon hydrophobic interactions between the protein sample and the adsorbent. However, unlike reversed phase supports, the bonded phases utilized in this method contain lower ligand densities and shorter ligand lengths (e.g. phenyl and butyl groups). As a result, surface hydrophobicity is reduced allowing bound proteins to be eluted under mild conditions, i.e. neutral pH and high ionic strength. Proteins initially bind to the support in the presence of high concentrations of salt (e.g., 2 to 4 M ammonium sulfate). The mechanism of binding depends upon the fact that many proteins, because of folding constraints caused by a variety of factors, such as disulfide bridging and glycosylation sites, contain hydrophobic areas on their surfaces. Under conditions which favor hydrophobic interactions, such as high salt concentrations found during a salting-out procedure, proteins will bind to nonpolar supports by way of these hydrophobic surface patches. During a gradient of decreasing salt concentration the bound proteins will then be differentially desorbed from the adsorbent according to their increasing hydrophobicities. Addition of an organic modifier such as methanol, ethanol or ethylene glycol to the eluting buffer increases its hydrophobicity and facilitates desorption off the column.

Although the underlying forces which determine the retention of proteins by hydrophobic interaction HPLC are recognized, understanding of all the effects which are involved is not complete. For example, it has been observed that protein retention on a hydrophobic interaction column is more sensitive to ligand length than is the case with reversed phase chromatography [181]. This was attributed to the hypothesis that in addition to nonpolar regions on the protein surface, sequestered hydrophobic residues in "pockets" near the surface are avilable to interact with the support ligand. The availability of these "pockets" depends upon the ability of the protein to undergo small conformational changes short of complete denaturation making it difficult to predict the elution order of different proteins in a mixture.

On a theoretical basis, proteins are likely to exhibit greater differences in their properties at their surfaces while in a native, globular form compared to a denatured form [182]. This accounts for why hydrophobic interaction chromatography is often reported to be equal to or in some cases superior to reversed phase chromatography in its ability to resolve protein mixtures [85]. The list of proteins and polypeptides that have been successfully purified by hydrophobic interaction chromatography is equally impressive as that indicated above for reversed phase HPLC. However, two notable examples should be emphasized. Because of its less disruptive elution conditions than reversed phase chromatography, it is particularly effective in retaining enzymatic activity [183,184] although enzymes that are inactivated by conditions of high salt concentrations are obvious exceptions [181]. Second, hydrophobic interaction HPLC has been used to purify proteins that undergo ion-dependent tertiary structural changes. α-Lactalbumin was isolated from milk whey by binding to a nonpolar support in the absence of calcium ions and eluting in an aqueous buffer containing Ca^{+2} ions [185]. Other proteins which undergo the opposite effect, i.e., a decrease in surface hydrophobicity in the absence of Ca^{+2} ions, have been purified by first binding to the column in the presence of Ca^{+2} ions and eluting in the presence of chelating agents [186-188]. Thus, one can readily appreciate the potential applications that hydrophobic interaction chromatography offers to the protein chemistry laboratory.

8.6 Affinity Chromatography

Of all the chromatographic techniques available, affinity-based methodologies are the most versatile and in some respects the most powerful. Similar to ion exchange, reversed phase and hydrophobic interaction chromatography, affinity chromatography is founded on the principle of constituents being adsorbed to and subsequently eluted from a solid support. The unique property that this technique offers is that the initial binding of the constituent is to a conformationally complementary surface which confers a level of specificity unparalleled by other methods. Affinity elution, like elution of the enzyme by the substrate, can leave all impurities behind and elute the enzyme alone off the column.

Specific interaction between a ligand and macromolecule is a universal feature of biological systems. In the realm of protein chemistry, these interactions are found between an enzyme and its substrate, inhibitor and cofactor, an antibody and antigen, a lectin and glycoprotein, as well as a hormone or vitamin and its receptor. In general, two types of affinity-based separations can be distinguished. Individualized interactions occur between one specific ligand and one protein as, for example, the reaction of a monoclonal antibody and its immunogen. In addition to these specific types of bindings between single substances, group specific adsorbents have been prepared that possess affinity for a certain class of related substances. These allow several substances with affinities for a given ligand to be purified without the need to prepare a different adsorbent for each. Thus, an immobilized lectin can be used to isolate a variety of different glycoproteins that possess similar carbohydrate moieties on different polypeptide backbones.

Although certain biological interactions had been exploited prior to the late 1960's in attempts to purify specific proteins, affinity chromatographic techniques did not become popular until after this time. Axen et al. [189] reported that carbohydrate-based gels could be modified with cyanogen bromide to provide a matrix for covalent coupling to primary amino groups of proteins. Since this initial observation, the list of different types of matrices and coupling agents that can be utilized to prepare an insoluble ligand support has grown tremendously. However, unlike the other methods of chromatography that have been successfully adapted to HPLC, affinity chromatography is only recently finding application in this area. This is in large part due to the fact that the advantage of high resolution offered by HPLC-based separations may not be needed if affinity chromatography in a traditional, low pressure system is by itself able to achieve purification to homogeneity in a single, high-yield step. For example, Whiting and Lindstrom [190] reported a 21,000-fold purification of a nicotinic acetylcholine receptor protein by a single-step immunoaffinity chromatographic procedure. Contaminants that are present following affinity chromatography are generally few and in such low concentration that if the sample is not already sufficiently pure, only one additional step (e.g. HPLC-based) may be necessary. Robb and Kutry [191] purified the human interleukin 2 receptor α chain by a two-step protocol utilizing immunoaffinity chromatography followed by reversed phase HPLC to remove a small percentage of remaining contaminants. Another limitation that has hindered the transition of affinity chromatography from conventional to HPLC chromatography is that silica-based supports may be incompatible with elution conditions commonly employed to desorb proteins. Silica particles are solubilized at high pH conditions [9] thus preventing their use in those cases where alkaline buffers have been found to be most effective in eluting proteins from immunoaffinity columns [192,193]. Figure 11 documents the immunoaffinity purification of an enzymatically active protein using 0.2 M NH_4OH, pH 11.0 for desorption.

The affinity HPLC column development that has occurred has been limited to analytical and small-scale preparative applications. Recently, however, the commercial availability of various preactivated HPLC affinity supports now makes this technique practical. Indeed, it does offer one distinct advantage unavailable with conventional, low-pressure affinity chromatography, i.e. its ability to ensure rapid neutralization of acidic or basic eluents thereby contributing to improved recovery of desorbed components [194]. Future developments in HPLC affinity chromatography in the industrial setting are particularly promising. In this regard, commercial application of HPLC affinity chromatography should prove superior to other separation techniques in its ability to separate biologically active proteins generated by recombinant DNA technology. Since eukaryotic proteins synthesized in bacteria normally must undergo correct folding and disulfide bridging to form fully active species and this process is often only fractionally efficient, HPLC affinity chromatographic methods utilizing active site ligands that only bind to target proteins containing a properly folded catalytic site can be used to separate active forms of a protein from its inactive isomers [195,196].

From the above discussion, it can readily be appreciated why affinity chromatography is presently more popular in the basic research laboratory rather than industrial laboratory. A comprehensive publication describing the theoretical and practical aspects of this method compiled over 200 references [197]. A few selected examples of these include the following: (1) purification of proteases such as elastase on a synthetic peptide substrate support [198] as well as esterase B [199] and cathepsin D [200] on solid supports of their respective competetive inhibitor, (2) isolation of DNA-binding proteins [201] and enzymes [202,203] on immobilized DNA, (3) use of carbohydrate supports for the purification of lectins [204,205] and glycosidases [206], (4) isolation of hormone-binding proteins and receptors utilizing affinity supports containing the specific hormone [207,208] or its selective antagonist [209], (5) purification of metalloproteins by metal ion chelate affinity chromatography [210,211] and (6) probing the surface topography of electron donor residues (cysteine, histidine, tryptophan) of enzymes by metal ion affinity chromatography [212].

Rather than review all realized and potential applications of affinity chromatography to protein purification, the remainder of this section will briefly emphasize two selected topics. The first is the use of antisera against synthetic peptides as solid

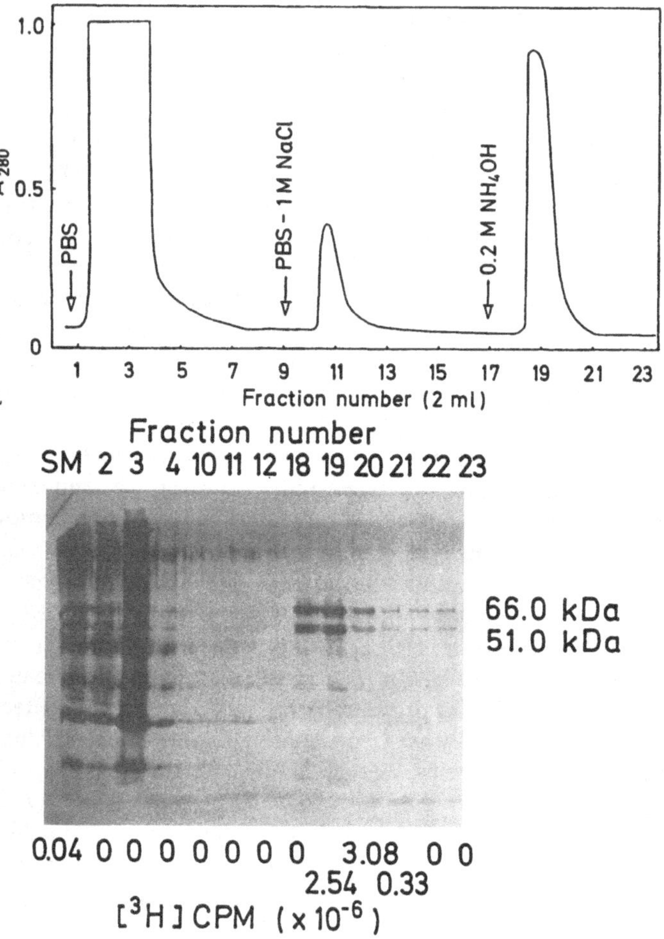

a

b

Fig. 11 a, b. Enzyme purification by immunoaffinity chromatography. A monoclonal antibody specific for the DNA-dependent RNA polymerase of human immunodeficiency virus type 1 (HIV-1) was coupled to CNBr activated Sepharose 4B, packed into a conventional low pressure column (2.5 × 2.0 cm) and used for the affinity purification of enzymatically active protein from a detergent extract of human lymphocytes infected with the virus. (**a**) illustrates the elution profile of unbound protein with phosphate buffered saline (PBS), pH 7.2, nonspecifically bound protein with PBS containing 1 M NaCl and specifically bound protein with 0.2 M NH_4OH, pH 11.0 (**b**) illustrates analysis of the starting material (SM) and selected elution fractions by immunoblotting with an anti-HIV-1 specific antiserum. The immunoblotting procedure involves separation of proteins by SDS-PAGE, electrotransfer of the proteins to nitrocellulose and visualization by reaction with an antibody. Specifically adsorbed enzyme was eluted beginning at fraction 19 and appeared as two bands of 51.0 and 66.0 kDa, the samller of which arises as a result of proteolytic cleavage of the C-terminus of the larger species. Enzyme activity, measured by the incorporation of [3H]thymidine into trichloroacetic acid insoluble RNA, is indicated below the Fig. and precisely correlates with the elution of antibody reactive protein. Courtesy of Dr. Robert Kilkuskie, Biotech Research Laboratories

supports for the purification of proteins from which the amino acid sequence of the peptide was deduced. The advent of recombinant DNA technology has resulted in the isolation and sequence analysis of a large number of genes for which either no known protein products exist or if they do exist, the proteins have not been purified. In many cases, these proteins can be identified and purified by affinity chromatography using antipeptide antibodies coupled to a solid support to adsorb the native protein followed by elution with the soluble peptide under mild conditions [213]. An additional step often involves a preliminary purification of the antipeptide antibodies on a peptide immunoadsorbent. These can then be eluted under acidic conditions [214] or more gently with the free peptide [215], the peptide removed by dialysis and the purified antibodies subsequently coupled to agarose to prepare an antipeptide affinity column for purification of the native protein. This method is fast and has been observed to produce a 2,500-fold purification in a single step [216]. Furthermore, it is a theoretically and technically feasible process to perform with the most difficult step being the selection of the appopriate amino acid sequence of the protein that will produce antibodies that react with the native protein. Several papers have summarized the important physicochemical properties of amino acids that should be considered when selecting a peptide sequence for immunization [217–219]. The potential application of antipeptide immunoadsorbents to protein purification is enormous particularly in those cases where only small amounts of the protein exist.

The second area in which affinity chromatography has produced a tremendous impact is the purification of antibodies, particularly monoclonal antibodies, by use of *Staphylococcus aureus* protein A and more recently group G *Streptococcus* protein G. Both proteins bind to immunoglobulins, the difference between them being the species and isotype specificity of this interaction [220,221]. Protein A binds to human, rabbit, mouse and guinea pig immunoglobulins while protein G binds to all these as well as those from the goat, sheep, cow and horse. The latter also interacts with human IgG3 and mouse IgG1, two isotypes that possess very weak affinities for protein A. The affinities which both proteins A and G possess for the isotypes to which they do bind are very high. Therefore, the conditions necessary to dissociate these complexes are relatively harsh, for example buffers of low or high pH which may be strong enough to partially denature the immunoglobulin. A recent analysis of the effects of various physical treatments on several different monoclonal antibodies revealed that a 30 min incubation at either pH 3.0 or 10.0 resulted in a 25% to 65% reduction in antibody titer [222]. As an alternative method, the silica-based, mixed-mode antibody exchanger (ABx) liquid chromatography ion exchange column [223] has been shown to produce a single-step purification of monoclonal antibodies under mild conditions and is particularly useful for those isotypes that bind weakly to protein A including murine IgG1 [224] and IgM [225]. The specificities of both proteins A and G make them ideally suited to purify monoclonal antibodies from either tissue culture supernatants or ascites fluids. Both bind to the Fc portion of immunoglobulins and thus they are useful as immunoadsorbents not only for purification of immune complexes without disruption of the antigen-antibody interaction [226] but also preparation of derivatives and fragments of IgG [227]. In the former case, protein A immunoadsorbents have been utilized to isolate immune complexes between monoclonal antibodies and their corresponding radiolabeled antigens. Once dissociated, the antigens can be isolated in sufficient purity for radiochemical structural studies by tryptic peptide mapping [228] or amino acid, sequence analysis [229]. In these examples, since the objective was a primary structural study of the purified proteins, strong dissociating conditions could be used to separate the antigen from the immunoadsorbent. However, in cases where conformation or biological activity must be preserved, milder elution conditions must be utilized. Wilchek et al. [31] have summarized various conditions that have been employed to elute proteins from immunoaffinity columns. These include low and high pH, urea and chaotropic salts. The effect that these agents will have on protein activity must always be considered prior to any attempt at immunoaffinity chromatography.

9 Optimizing Preparative-Scale Isolations

The development of a scale-up strategy follows several well-defined guidelines based upon the amount of sample involved, the particle size of the adsorbent and column dimensions [230]. These are summarized in Table 3. First, the purification steps are

Table 3. Step-by-step optimization and scale-up strategy

Step	Sample amount	Particle size	Column dimensions
Optimization of chromatographic conditions	1–10 mg	5–10 μm	3–8 mm D 7.5–30 cm L
Comparison of analytical and preparative particles	1–10 mg	50–100 μm	3–8 mm D 7.5–30 cm L
Determination of maximum mass load	10–100 mg	50–100 μm	3–8 mm D 7.5–30 cm L
Scale-up	1–100 g	50–100 μm	5–15 cm D 20–50 cm L

D, diameter; L, length

optimized on an analytical scale to achieve the maximum degree of purity of the protein of interest. With regards to HPLC techniques, such micropreparative separations are generally performed with less than 10 mg of sample using columns with internal diameters of 3 to 8 mm and lengths of 7.5 to 20 cm and packed with particles of sizes 5 to 10 μm. Parameters that can be changed to achieve optimum resolution include gradient conditions, strength of organic solvent, flow rate and temperature. Next, the analytical separation is repeated using an identical sample size and the same analytical column but packed with particles to be used in the preparative-scale isolation, for example those with sizes of 50 to 100 μm. This allows one to confirm that the separation parameters achieved with the analytical and preparative-size packings are similar, particularly with respect to column selectivity. The third step in the scale-up process is to increase the sample size and determine the mass load level of the preparative-size particles in the analytical column. This is achieved by resolving progressively larger amounts of sample and determining the maximum load that can be accommodated without surpassing the binding capacity of the column. Finally, the separation is directly transferred to a large preparative-scale column containing preparative-size particles.

The chromatographic elution conditions required for the process-scale separation are calculated based upon the respective dimensions of the analytical and preparative columns. For example, determination of the preparative flow rate (FRp) is made with the formula

$$\mathrm{FRp} = [(\mathrm{Dp}^2/\mathrm{Da}^2) \times \mathrm{FRa}]\,, \tag{3}$$

where Dp is the diameter of the preparative column, Da is the diameter of the analytical column and FRa is the flow rate of the analytical column. Both FRp and FRa are related to their respective linear velocities (LV) which are experimentally determined with the equation

$$\mathrm{LV} = \mathrm{L}/\mathrm{Rt}\,, \tag{4}$$

where L is the column length and Rt is the retention time of an unretained peak such as acetone. From a plot of FRa vs. LVa and FRp vs. LVp, the actual value of FRp that equates LVp and LVa at the optimal value of FRa can then be deduced. Similarly, the scale-up factor (SF) used to determine the maximum mass load that can be accommodated when changing from an analytical to preparative-scale column is determined as

$$\mathrm{SF} = (\mathrm{Dp}^2\mathrm{Lp})/(\mathrm{Da}^2\mathrm{La})\,, \tag{5}$$

where Lp and La are the lengths of the preparative and analytical columns, respectively. A small-scale separation of 10 mg of protein on a 0.8×10 cm column can therefore be increased by a factor of 152 to 1.5 g on a 5.7×30 cm column.

10 Descriptions of Electrophoretic Techniques

Separation of proteins and polypeptides by electrophoresis has been the standard method of following a purification protocol and analyzing the homogeneity of the final product. However, the great resolving power exhibited by some of the more recently introduced modifications has also led to the development of preparative-scale electrophoretic techniques. Generally, proteins are electrophoresed through a support matrix, commonly paper, cellulose acetate, starch, agarose or polyacrylamide gel. Their rate of migration depends on the strength of the electrical field, the net charge, size and shape of the molecules as well as the ionic strength, viscosity and temperature of the matrix. Zonal electrophoretic methods utilize a buffer with a pH above the pI of the proteins in the sample and separations are based solely on net charge. Since agarose and polyacrylamide gels provide a porous matrix, these media will separate by charge and molecular weight. Agarose, a highly purified polysaccharide derived from agar, is not commonly used in protein electrophoresis because of its poor sieving (resolving) properties and fragility. Polyacrylamide is formed by the polymerization of acrylamide monomers that are covalently crosslinked and has become the matrix of choice for resolving protein mixtures on an analytical scale. The major limitation that has hampered the scale-up of PAGE has been the relatively small amount of material that can be accommodated by the standard apparatus. As detection limits of peptides and proteins are continually reduced, however, PAGE is increasingly becoming a powerful tool to achieve their purification, particularly for primary structural investigations. Initially, proteins in polyacrylamide gels were visualized by staining with Coomassie blue which is sensitive to about 1 μg of protein. Newer staining techniques utilizing silver or nickel [231,232] provide sensitivity to approximately 10 ng of protein. The most sensitive detection is produced by autoradiography where radiolabeled proteins are visualized after placing the gel in contact with x-ray film. The remainder of this section will briefly summarize recent advances in PAGE, i.e. SDS-PAGE, isoelectric focusing and two dimensional PAGE.

10.1 Sodium Dodecyl Sulfate Polyacrylamide Gel Electrophoresis (SDS-PAGE)

SDS, an anionic detergent, binds to proteins in a manner that overwhelms their individual charge and induces a conformational change to rod-like structures. As a result, all proteins possess an equal charge density per unit length such that electrophoretic migration is determined by molecular weight. SDS-PAGE is performed using either a continuous buffer system [233] or, more commonly, under discontinuous conditions such as those described by Laemmli [234], Neville [235] and Wyckoff et al. [236]. The latter case is portrayed in Fig. 12. A stacking gel containing a large pore po lyacrylamide overlays a separating gel with a more restrictive pore size. Both gels

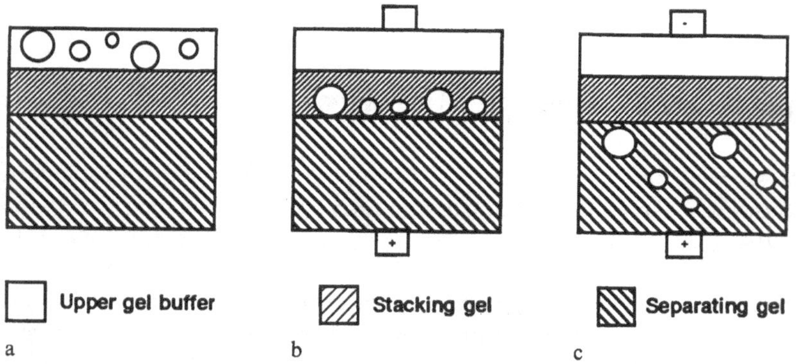

Fig. 12a–c. The principle of discontinuous electrophoresis in the presence of SDS. **a.** Proteins are treated with a reducing agent and SDS to dissociate multimers into their constituent polypeptide chains with similar charge to mass ratios. **b.** When an electric field is applied, the polypeptides migrate into the stacking gel between two different pH buffers forming a sharp zone at the interface of the separating gel. **c.** The polypeptide chains enter the separating gel where they are resolved according to their molecular weight

Table 4. Useful protein molecular weight standards

Protein (source)	Approximate molecular weight (Da)	Supplier
Blue dextran 2000	2,000,000	PL, S
Thyroglobulin (cow)	670,000	B, S
Urease, hexamer (jack bean)	575,000	S
Ferritin (horse)	440,000	C, P, PL, S
β-Glucoronidase	290,000	C
Urease, trimer (jack bean)	272,000	S
Catalase (cow)	232,000	P, PL
β-Amylase (sweet potatoe)	200,000	S
Myosin (rabbit)	200,000	B, D, L, S
α_2-Macroglobulin (human)	180,000	S
RNA polymerase, β′-subunit	165,000	P
Aldolase (rabbit)	158,000	C, P, PL
γ-Globulin (cow)	158,000	B
RNA polymerase, β-subunit	155,000	P
β-Galactosidase (*Escherichia coli*)	116,000	B, D, S
Hexokinase	100,000	C
Phosphorylase B (rabbit)	97,400	B, D, L, S
Fructose-6-phosphate kinase (rabbit)	84,000	S
Ovotransferrin (chicken)	78,000	PL
Hemocyanin (*Limulus polyphemus*)	70,000	S
Serum albumin (cow)	66,200	A, B, C, D, L, P, PL, S
Pyruvate kinase (chicken)	58,000	S
Glutamate dehydrogenase	55,000	D
Fumarase (pig)	48,500	S
Ovalbumin (chicken)	45,000	A, B, C, D, L, P, PL, S
Alchohol dehydrogenase	39,000	D
RNA polymerase, α′-subunit	39,000	P
Glyceraldehyde-2-phosphate dehydrogenase (rabbit)	36,000	S
Lactate dehydrogenase (rabbit)	36,000	D, S
Pepsin	34,700	C, S
Carbonic anhydrase (cow)	30,000	B, C, D, S
Triosephosphate isomerase (rabbit)	26,600	S
Chymotrypsinogen A (cow)	25,000	C, L, P, PL
Trypsin inhibitor (soybean)	21,500	A, B, D, P, S
β-Lactoglobulin B (cow)	18,400	L, S
Myoglobulin (horse)	17,200	B, D
Hemoglobin (cow)	16,000	S
Lysozyme	14,400	B, C, L, S
α-Lactalbumin (cow)	14,200	S
Ribonuclease A	13,700	PL
Cytochrome C (horse)	12,400	A, C, D, P, S
Aprotinin (cow)	6,500	A, S
Trypsin inhibitor (cow)	6,200	L
Insulin, alpha and beta chains	3,000	A, D, L

Table 4. (continued)

A: Amersham Corperation, Arlington Heights, IL;
B: BioRad Laboratories, Richmond, CA;
C: Calbiochem, San Diego, CA;
D: Diversified Biotech, Newton Centre, MA;
L: Life Technologies, Inc, Gaithersburg, MD;
P: Pierce Chemical Company, Rockford, IL;
PL: Pharmacia LKB Biotechnology, Inc, Piscataway, NJ;
S: Sigma Chemical Company, St. Louis, MO

as well as the electrophoresis running buffer possess different pH and ion concentrations. As proteins migrate through the stacking gel, they are concentrated in a thin zone between the leading buffer ion of the stacking gel and the trailing ion of the running buffer. Upon reaching the separating gel, protein mobility becomes inversely proportional to molecular weight. Using a set of predetermined molecular weight markers (Table 4) to construct a standard curve of protein mobility, the apparent molecular weight of an unknown can then be calculated. Glycoproteins containing a high percentage of carbohydrate are an exception since they bind to smaller amounts of SDS on a weight basis and thus exhibit reduced gel mobility.

As an analytical tool, discontinuous SDS-PAGE offers a very high degree of resolution necessary to follow a purification scheme. Optimization of the running conditions has permitted sharp resolution of proteins over a molecular weight range of 1,500 to 100,000 Da on a single gel [237]. Identification of the molecular weight of impurities that may contaminate the protein of interest allows one to choose an appropriate liquid chromatographic technique to achieve a greater degree of purity. In most instances, however, the very component which makes this useful for analyzing protein homogeneity (i.e. SDS) also renders it impractical as a preparative technique. The notable exception to this limitation applies when the goal of protein purification is a primary structural rather than functional analysis. Proteins resolved on SDS containing polyacrylamide gels can often be recovered by electroelution for direct amino acid sequence analysis via amino-terminal Edman degradation (see below).

10.2 Isotachophoresis

The principle of electrophoretic separation by isotachophoresis is very similar in many respects to that of disc electrophoresis discussed above, in particular the effect of sample ion migration through the stacking gel. Thus, sample components migrate as a zone in an aqueous medium between a leading buffer ion and a trailing ion and are separated from each other in order of their charge differences [239]. Since the leading and trailing ions are chosen such that the sample ions have an intermediate mobility between them, a zone sharpening effect tends to conteract the effects of diffusion. Furthermore, addition of intermediate mobility compounds, such as amino acids, to define the boundaries of the individual sample components has resulted in the resolution of isotachophoresis being in many cases equivalent to that of other electrophoretic techniques. As an analytical technique, isotachophoresis provides the additional benefits of small sample requirements and short analysis time. On the preparative-scale, however, its full potential is yet to be realized. Incorporation of on-line detection and fraction collection systems has, in a few cases, enabled isotachophoresis to be utilized as an integral step in protein purification procedures [239]. The reader is referred to the review article by Holloway and Battersby [240] for further reference.

10.3 Isoelectric Focusing

The migration of proteins in isoelectric focusing (IEF) gels is determined by their pI and the development of a pH gradient which is formed by inclusion of highly mobile amphoteric components (ampholytes) in the gel matrix (Fig. 13). Ampholytes consist

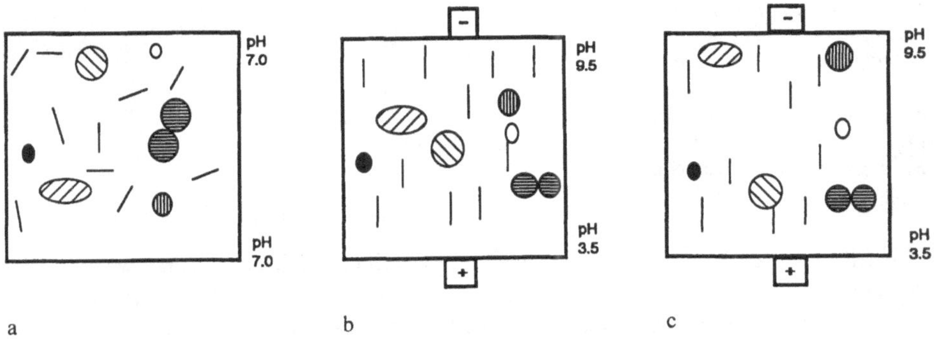

Fig. 13a–c. The principle of isoelectric focusing. **a.** Proteins are applied to a gel containing carrier ampholytes (indicated by the *small bars*). A random distribution of ampholytes produces a uniform pH of 7.0 throughout the gel. **b.** When an electric field is applied, the ampholytes quickly migrate to their respective isoelectric points forming a pH gradient, illustrated here as extending from pH 3.5 to 9.5. The protein molecules migrate slowly to the pH corresponding to their isoelectric points. **c.** An equilibrium is reached when the net charge of the proteins is 0 and their migration stops

of a collection of zwitterionic, small molecular weight polyaminopolycarboxylic acids with differing pIs. In the presence of an electric field, they arrange themselves in a manner such that the most basic are located near the cathode and the acidic near the anode. A focusing effect is achieved when the proteins migrate to the position in the pH gradient where their net charge is zero. Table 5 lists several useful protein isoelectric point standards that can be utilized to estimate the pI of an unknown protein. IEF is primarily a tool for analytical purposes although it can be used for preparative-scale isolations using either a column apparatus with a sucrose gradient to minimize thermal-induced convection currents [241] or a horizontal electrophoretic chamber [242]. High resolution preparative IEF has recently been reviewed and this paper can be consulted for further information [243]. However, a clear disadvantage relating to the use of IEF in a preparative mode is the tendency of proteins to precipitate out of solution at their pI where they are less soluble than in a charged state.

Table 5. Useful protein isoelectric point standards

Protein (source)	Approximate pI	Supplier
Cytochrome C (horse)	10.60	B, PL
Trypsinogen (cow)	9.30	S
α-Chymotrypsin	8.80	B
Lactate dehydrogenase (rabbit)	8.30–8.55	S
Myoglobin (whale)	8.05	B
Myoglobin (horse)	6.76–7.16	B, S
Carbonic anhydrase (human)	6.50	B, S
Carbonic anhydrase (cow)	5.85–6.00	B, S
Azurin (*Pseudomonas aeruginosa*)	5.65	PL
β-Lactoglobulin A (cow)	5.13–5.25	PL, S
β-Lactoglobulin B (cow)	5.10–5.35	B, PL
Phycocyanin	4.65	B
Trypsin inhibitor (soybean)	4.55	S
Glucose oxidase (fungal)	4.25	PL
Amyloglucosidase (*Aspergillus*)	3.55–3.65	PL, S

B: BioRad Laboratories, Richmond, CA;
PL: Pharmacia LKB Biotechnology, Inc, Piscataway, NJ;
S: Sigma Chemical Company, St. Louis, MO

Since mobility in analytical IEF gels is not influenced by molecular weight, it is desirable to utilize a matrix containing nonrestrictive pores. As a consequence, both agarose [244], polyacrylamide gels [245] and granulated gels of the Sephadex or BioGel type [246] have proven to be popular choices. The resolving potential of IEF is tremendous. Proteins with a single charge difference can be separated on an analytical scale [247]. In practical terms, this translates into a difference in pI values of 0.001 to 0.01. In many cases this is a clear advantage over SDS-PAGE allowing the detection of impurities following size exclusion chromatography, for example. It must be cautioned, however, that charge heterogeneity of proteins can also be produced by post-translational modifications including phosphorylation [248] and glycosylation [247]. A certain degree of charge heterogeneity due to these effects may offen be tolerated in a protein preparation when the objective of the purification protocol is, for example, to prepare an antigen for immunization or primary structural analysis [249].

10.4 Two-Dimensional Polyacrylamide Gel Electrophoresis

Both SDS-PAGE and IEF are each capable of resolving 30 to 40 different proteins on a single gel. Resolution of between 100 and 200 components, however, can be achieved by sequential combination of these techniques [40]. Commonly, IEF is performed first in a small diameter tube gel which is subsequently layered over a SDS-PAGE slab gel [250]. This principle is illustrated in Fig. 14. Separations in the reverse order have also been reported, i.e. SDS-PAGE preceeding IEF [251]. In either case, the handicaps of both SDS-PAGE and IEF when used on a preparative-scale are apparent

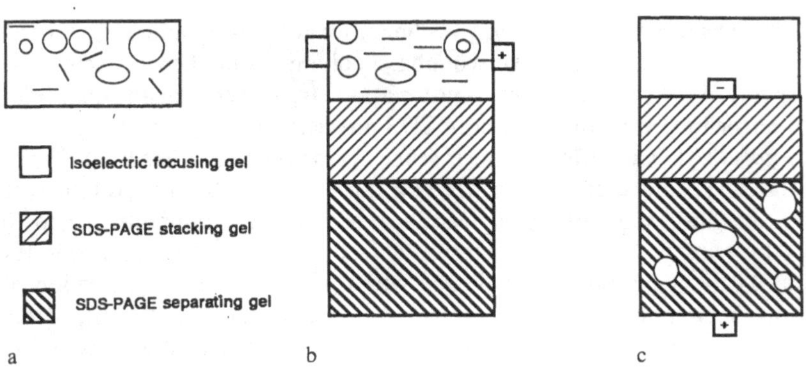

Fig. 14a–c. The principle of two dimensional gel electrophoresis. **a.** Proteins are applied to an isoelectric focusing gel. Carrier ampholytes are indicated by the small bars. **b.** The proteins are first separated according to their isoelectric points. The isoelectric focusing gel is then connected to an SDS-PAGE gel at a right angle to the original separation. **c.** The polypeptide chains are finally resolved in the separating gel according to their molecular weight

in two-dimensional PAGE. The great utility of this technique lies in its application to monitoring a purification scheme. For instance, identification of a contaminant with the same pI but different molecular weight as the protein of interest directs one to use ultrafiltration or size exclusion chromatography as the next purification step. In contrast, if the impurity possesses a different pI but similar molecular weight, ion exchange chromatography can be used at the next step. A contaminant of similar charge and molecular weight as the protein of interest suggests the use of a hydrophobicity-based separation step, such as reversed phase or hydrophobic interaction chromatography. Like SDS-PAGE, two-dimensional PAGE is limited in its applicability to isolation of proteins for amino acid sequence analysis and antibody production. The former was discussed above. Both monoclonal [252] and polyclonal [253] antibodies have been prepared using proteins excised from Coomassie blue or silver stained polyacrylamide gels. Since such proteins are denatured, however, the antibodies that are produced are frequently directed against linear rather than conformational determinants. One of the best methods of detecting their specificity is the use of immunoblotting (Western blotting) procedures whereby proteins separated by one or two dimensional PAGE under denaturing conditions are electroblotted to nitrocellulose membranes and reacted with the antibodies [254].

11 Amino Acid Sequencing of Proteins

11.1 Micro-Sequence Analysis

Polyacrylamide gel electrophoresis in the presence of sodium dodecyl sulfate can resolve complex mixtures of polypeptides. Electrophoretic elution from gel slices or electrophoretic transfer to nitrocellulose or glass fiber membranes is used to recover small quantities of proteins from SDS-PAGE gels for peptide mapping and protein sequencing. Proteins can be resolved on two-dimensional polyacrylamide gels, pressure extracted onto short HPLC reversed phase columns and then collected for direct sequencing [73,255–257]. Sequence information from very small quantities of proteins electroeluted from discontinuous SDS polyacrylamide gels [258] or electroblotted onto polyvinylidene difluoride membrane [259] or on glass fiber filter [255,260] is routinely obtained and used for synthesizing nucleotide probes for isolation of genes. Current methods of gas phase sequence analysis require a minimum of 5 to 10 picomoles of protein [261]. For an average protein of 10,000 to 50,000 Da, this represents from 0.05 to 0.5 µg of material, well within the sensitivity of detection of proteins in polyacrylamide gels using present staining technology. Samples from HPLC separations are collected directly. By optimizing the instrument and reduction of column internal diameter, detection can be improved more than tenfold [262]. Future developments in amino acid sequencing are predicted to decrease the amount

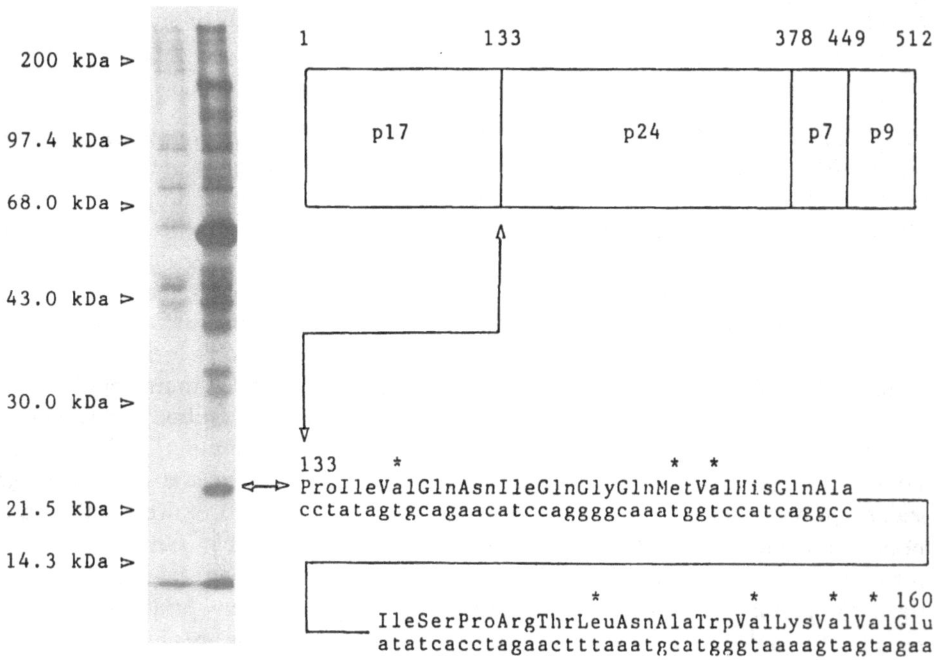

Fig. 15. Purification and N-terminal amino acid sequence analysis of a viral protein derived from a polyprotein precursor. Human lymphocytes were infected in vitro with the human immunodeficiency virus type 1 (HIV-1), cultured in medium containing [35S]methionine plus [3H]valine or [3H]leucine and lysed in detergent solution. Viral proteins were recovered by immunoprecipitation, resolved by SDS-PAGE and visualized by fluorography. Specific immunoprecipitation of viral proteins was determined by comparison of those recovered using a normal human serum negative control (left lane) with those precipitated by an antiserum from an individual infected with HIV-1 (right lane). The migration of molecular weight markers is indicated on the left in kDa. A protein of 24,000 Da (indicated by *arrow*) was cut out of the gel, recovered by electroelution and its amino acid sequence determined (*lower right*). Based upon the radioactive amino acids identified (indicated by *asterisk*), the N-terminus of this protein (designated p24) was localized to residue 133 of the 512 amino acid open reading frame previously demarcated by DNA sequencing (*upper right*). The DNA sequence of this segment is given in lower case letters below the amino acid sequence. The N-termini of the p7 and p9 proteins isolated on other SDS gels were similarly localized to residues 378 and 449, respectively

of protein required to the attomole range[263] enabling SDS-PAGE, particularly in a two dimensional format, to become an extremely powerful protein purification technique.

11.2 Radiochemical Sequence Analysis

One modification of this procedure that is currently available to achieve a comparable level of sensitivity is the use of radioactively labeled amino acids in the purification and sequence analysis of proteins synthesized by cells in tissue culture. [35S]-labeled methionine or cysteine either alone or in combination with [3H]-labeled amino acids are metabolically incorporated into cellular or viral proteins which can subsequently be resolved on one or two dimensional SDS gels, isolated by electroelution and directly analyzed by N-terminal Edman degradation[264-266]. Figure 15 illustrates how this technique has been utilized to identify natural proteolytic cleavage sites in viral polyproteins whose DNA sequences have previously been established.

12 Recombinant DNA in Protein Overproduction

12.1 Prokaryotic Expression

As a result of the advent of recombinant DNA technology the list of genes for numerous protein products of pharmaceutical and industrial importance is growing rapidly. The technology for cloning and sequencing of genes is better perfected than purifying and characterizing their protein products. Most proteins can be purified in microgram amounts by using fast protein liquid chromatography, high-performance liquid chromatography, affinity chromatography or elution from SDS polyacrylamide gels and renaturation. However, some proteins need to be produced in kilogram quantities and homogeneous purity to meet industrial needs. Expression of cloned genes was originally developed in bacterial systems but has since been accomplished in insect and eukaryotic host cells as well. Plasmids pKc30 and pKc31 have been used for large-scale bacterial expression of cloned genes[267]. In plasmid pKc31, the pL promotor originally from phage is kept repressed by a thermolabile repressor, CITS 857. The cells are usually grown to mid-log phase and then heat is used to induce the synthesis of the cloned protein under the control of the pL promotor [268]. Construction of an overexpressing strain permitted for the first time purification to homogeneity of chorismate synthase from *E. coli*[269].

12.2 Eukaryotic Expression

A major disadvantage in using bacterial host cells for expression of recombinant proteins concerns the fact that prokaryotic translation systems do not recognize eukaryotic post-translational signals such as sites of glycosylation, phosphorylation and signal peptide cleavage. Insect and eukaryotic host expression systems, in contrast, produce recombinant proteins containing such post-translational modifications. The recombinant baculovirus vector has been the most widely utilized for expression of cloned genes in insect cells[270,271]. Recently, a protein-free medium that is capable of supporting large-scale insect cell culture, virus propagation and cloned gene expression has been developed to enhance the purification of recombinant proteins [272]. Eukaryotic viruses have been modified to produce vectors for cloning, selection and expression of heterologous genes. Examples of these include vaccinia virus, simian virus 40, adenovirus and a variety of retroviruses. Proper post-translational modifications and processing of recombinant proteins produced in these systems has been documented [273]. Several review articles on this subject have appeared and may be consulted for further reference [274–276].

Rational alteration of enzyme properties using site-directed mutagenesis of cloned genes is being examined using a variety of model systems and these have been recently summarized [277]. Single amino acid substitutions have produced mutant enzymes from *Bacillus amyloliquefaciens* with increased activity, altered specificity and altered pH activity [278]. Changing an amino acid in the active site of an enzyme also alters its substrate interactions. Amino acid substitutions in *B. thermoproteolyticus* subtilisin at positions 131 and 172 were shown to decrease the rate of thermal inactivation via a mechanism involving enhanced Ca^{+2} ion binding [279].

13 Fusion Proteins

13.1 ß-*Galacto*sidase Fusion Proteins

Several *E. coli* cloning vectors in which DNA fragments can be cloned into the β-galactosidase (lacZ) gene via restriction sites are avilable. The protein coded by the inserted DNA fragment is expressed as a chimeric β-galactosidase fusion protein (Fig. 16). This technique offers several advantages such as ease of screening of recombinants, ease of purification and enhanced stability of the foreign gene product in the host cells. A variety of promotors have been used to express the fusion proteins at significant levels. A popular double-stranded bacteriophage, lambda gt11, is used for inserting DNA near the 3′ (C terminus) end of the lacZ gene, 53 base pairs upstream from the translation termination codon [280,281]. Monoclonal antibodies which recognize native β-galactosidase as well as N- or C-terminal β-galactosidase fusion proteins are commercially available. These antibodies can be used to purify the recombinant proteins by immunoaffinity chromatography and the fusion proteins subsequently used for immunization to prepare an antiserum from which the antibodies against the foreign gene product can be isolated by passage through a β-galactosidase affinity column [282]. β-Galactosidase fusion proteins have also been purified by binding to and elution from aminophenylthiogalactosidyl-Sepharose columns [283].

13.2 Other Fusion Proteins

Several notable handicaps have been encountered with the production and purification of fusion proteins using the techniques described above. First, since the added protein segment is quite large, the foreign gene product often does not properly fold into a native, active conformation [284]. Denaturation followed by slow renatur-

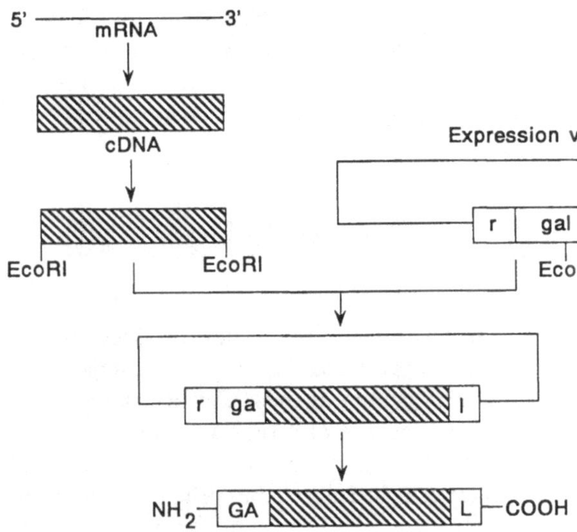

Fig. 16. Expression of a foreign gene as a β-galactosidase fusion protein. The foreign gene is first synthesized as cDNA from mRNA and suitable restriction endonuclease sites (e.g. EcoRI) generated at each end. The cDNA (hatched box) is then introduced into the homologous restriction endonuclease site in the β-galactosidase gene (lac Z) of the expression vector (open box). The lac Z gene is preceeded by the necessary regulatory DNA sequences required for gene expression, e.g. start codon (ATG), promotor (p), operator (o) and ribosome binding site (r). Note that the DNA sequence of the lac Z gene is dissected by the cDNA. Expression of the hybrid gene after transfection of bacterial host cells result in a fusion protein containing the cloned protein inserted into the β-galactosidase protein and terminated at the stop codon (TGA) of the cDNA or lac Z gene. Unlike the normal β-galactosidase protein, the fusion protein is unable to hydrolize synthetic substrates enabling selection of bacterial cells containing the fusion protein by a color reaction.

ation may be required to obtain a native state of the introduced polypeptide [285, 286]. Second, the high binding affinities of most monoclonal antibodies and polyclonal antisera against the host protein segment necessitates use of severe denaturants for elution from immunoaffinity columns. Finally, once the fusion protein has been purified, it has often been difficult to remove the added protein from the desired polypeptide. Attempts to circumvent this problem have employed chemical [286–289] or enzymatic [284, 290, 291] methods. Recently, the use of a short, hydrophilic peptide engineered into the N-terminus of fusion proteins has solved many of these problems [291]. This peptide was less than 10 amino acids in length and did not interfere with the folding and subsequent biological activity of the desired proteins. Furthermore, an antipeptide antiserum can be prepared for immunoaffinity purification using the free peptide to elute bound fusion protein under physiological conditions. Engineering the proper peptide sequence in the fusion protein allows its removal by proteolytic enzymes of appropriate specificity. Alternatively, use of a lysine-rich peptide has permitted biotinylation of the recombinant protein without loss of biological activity such that it may serve as a ligand for affinity purification of a variety of receptors on avidin-containing supports [292].

When proteins are overexpressed in *E. coli*, some of them accumulate as insoluble aggregates known as inclusion bodies. Sometimes the formation of inclusion bodies can be prevented by changing the media and growth conditions of *E. coli.* Proteins can also be engineered so that they are secreted out of the cell. However, many proteins are extracted from the cells and renatured to gain biological activity. The protein contained in inclusion bodies may be solubilized by treatment with chaotropic agents, acids or alcohols.

14 Future Developments

The future of protein purification rests on the development of new products as well as innovative technologies of microanalytical separations. In the former category are fluoropolymer supports for immobilization of affinity ligands. Covalent attachment of a perfluoralkylating chain to primary amino groups of proteins allows them to be strongly adsorbed to polyfluorocarbon-based supports in either a solid or liquid phase [293]. Continuous liquid-liquid extraction based upon this principle may soon become common industrial practice to purify recombinant proteins from spent media. A similar type of liquid-liquid extraction occurs in centrifugal partition chromatography, a new application based upon the distribution of solute mixtures between two liquid phases in the absence of a solid support [294]. Unlike traditional counter-current partition chromatography, this technique utilizes separation columns connected in series within cartridges arranged around a centrifuge rotor. The mobile phase, which is often the less dense of the two immiscible solvents, is pumped through the columns where, in droplet form, it comes in contact with the stationary phase which is maintained in the system by centrifugal force. Another novel application soon to become commercially popular is radial flow liquid chromatography. Instead of the conventional mode of applying a sample to the top of a column and separating it with downward axial flow, the sample is distributed around the outer circumference of the column and separated by radial flow (Fig. 17). The capacity of the column can therefore be increased by a factor proportional to the length of the colum. Capillary

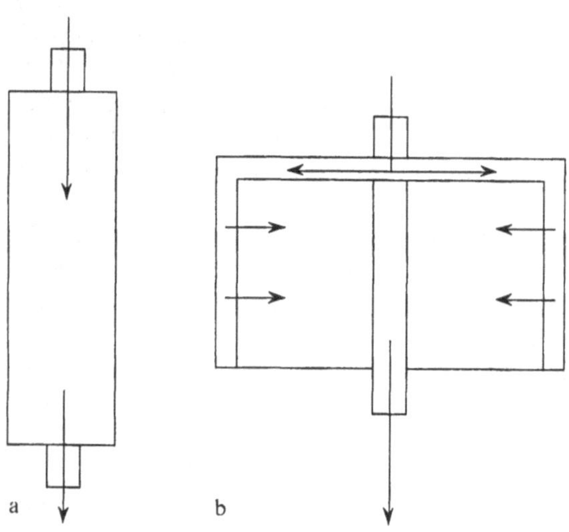

Fig. 17a, b. The principle of radial flow chromatography. **a**. Conventional tubular liquid chromatography with downward solvent flow. **b**. Radial flow chromatography employes solvent flow from the outer circumference of the column inward to a central collection tube

electrophoresis and open tubular liquid chromatography represents promising new methods of analytical protein separation utilizing sample sizes on the nanoliter scale [263]. Fused silica capillaries with internal diameters of less than 100 μm and wall thicknesses of less than 200 μm allow efficient dissipation of heat to permit extremely high resolution separations. Furthermore, the high degree of purity of silicone dioxide used in manufacturing these capillary tubes provides a very inert surface for improved chromatography of bioactive proteins. Capillary electrophoresis can be performed in a variety of different types of media, including aqueous or organic solutions or capillaries filled with agarose or polyacrylamide. A recent review has compiled a list of both standard and unknown proteins that have been separated by this method [295] although more sensitive detectors need to be developed before its full potential is realized. Open tubular liquid chromatography also utilizes long, thin capillaries but to which a stationary phase such as that used in reversed phase chromatography (e.g. C-18) has been coated onto the inner wall surface. Again, the major obstacle that needs to be overcome before this technique becomes more applicable to the protein chemistry laboratory is the development of sensitive methods of detection. Once this and other obstacles are overcome, future protein separations will be capable of analyzing single cell volumes on a routine basis.

15 References

1. Thompson H (1988) Am. Lab. *20*: 113
2. Scopes R (1982) Protein purification: principles and practice. Springer Verlag, Berlin Heidelberg New York
3. Verzele M, Dewaele C (1985) Preparative high performance liquid chromatography. A practical guideline. TEC, Gent, Belgium
4. Wankat PC (1986) Large-scale adsorption and chromatography. Vols 1 and 2, CRC, Boca Raton, FL
5. Burgess RR (1987) Protein purification. Alan R. Liss, New York, NY
6. Jakoby WB (1984) Enzyme purification and related techniques. Part C. In: Methods in enzymology, vol 104
7. Regnier FE (1983) Methods Enzymol. *91*: 137
8. Regnier FE (1983) Science *222*: 245
9. Gooding KM (1986) BioChromatogr. *1*: 34
10. Lottspeich F (1986) Chromatographia *22*: 407
11. Potter RL, Lewis RV (1986) High-Perform. Liq. Chromatogr. *4*: 1
12. Rahn P, Joyce W, Schratter P (1986) Am. Biotechnol. Lab. *4*: 34
13. Ratafia M, Keenan J (1986) Am. Biotechnol. Lab. *4*: 40
14. Freiser HH, Gooding KM (1987) BioChromatogr. *2*: 186
15. Guiochon G, Katti A (1987) Chromatographia *24*: 165
16. Hearn MTW (1987) J. Chromatogr. *418*: 3
17. Johnson RD (1987) Dev. Ind. Microbiol. *27*: 77
18. Huang JX, Guiochon G (1988) Biochromatogr. *3*: 140
19. Pearson JD, McCroskey MC, Dewald DB (1987) J. Chromatogr. *418*: 245
20. Smith JA, O'Hare MJ (1986) Methodol. Surv. Biochem. Anal. *16*: 17
21. Dean PDC, Walker JNB (1985) Biochem. Soc. Trans. *13*: 1055
22. Matsushita N (1985) Pharm. Technol. Jpn. *1*: 775
23. Bietz JA (1986) Adv. Cereal Sci. Technol. *8*: 105
24. Burnouf T, Bietz JA (1987) Seed Sci. Technol. *15*: 79
25. Heukeshoven J, Dernick R (1988) Chromatographia *25*: 230
26. Regnier FE (1984) Recept. Biochem. Methodol. *2*: 61

27. Roumeliotis R, Unger KK (1979) J. Chromatogr. *185*: 445
28. Unger K (1984) Methods Enzymol. *104*: 154
29. Wenzel HR, Beckmann J, Mehlich A, Siekmann J, Tschesche H (1985) Mod. Methods Protein Chem. *2*: 173
30. Hyder SM, Witliff JL (1987) Biochromatogr. *2*: 121
31. Wilchek M, Miron T, Kohn J (1984) Methods Enzymol. *104*: 3
32. Walters RR (1985) Anal. Chem. *57*: 1099A
33. Kasai K (1986) Tanpakushitsu Kakusan Koso *31*: 719
34. Carr PW, Bergold AF, Hanggi DA, Muller AJ (1987) Chromatogr. Forum *1*: 31
35. Sofer G, Britton VJ (1983) BioTechniques *1*: 198
36. Rappuoli R, Perugini M, Marsili I, Fabbiani S (1983) J. Chromatogr. *268*: 543
37. Wilson KJ (1985) An introduction to protein/peptide isolation strategies for primary structure determinations. Applied Biosystems User Bulletin, Issue 1, Applied Biosystems, Foster City, CA
38. Vardanis A (1985) J. Chromatogr. *350*: 299
39. Tomany M (1987) Higher purification on Waters PROTEIN-PAK DEAE 5 PW ion exchange column when loading the sample in a specified salt concentration. Lab Highlights # 0348, Millipore, Bedford, MA
40. Dunbar BS (1987) BioTechniques *5*: 218
41. Anderson NG, Anderson NL (1982) Clin. Chem. *28*: 739
42. Fox JA, Livingston DJ, Orme-Johnson WH, Walsh CT (1987) Biochemistry *26*: 4219
43. Calderon J, Sheehan KCF, Chance C, Thomas ML, Schreiber RD (1988) Proc. Natl. Acad. Sci. USA *85*: 4837
44. Palker TJ, Clark ME, Sarngadharan MG, Matthews TJ (1987) J. Virological Methods *18*: 243
45. Akerstrom B, Logdberg L, Babiker-Mohamed H, Lohmander S, Rask L (1987) Eur. J. Biochem. *170*: 143
46. Riss TL, Sirbasku DA (1987) In Vitro *23*: 841
47. Hashizume S, Kuroda K, Murakami H (1983) Biochim. Biophys. Acta *763*: 377
48. Rubin JS, Bradshaw RA (1984) Preparation of guinea pig prostrate epidermal growth factor. In: Barnes RA, Sirbasku DA, Sato GH (eds) Methods for preparation of media, supplements, and substrata for serum-free animal cell cultures. Alan Liss, New York, p 139
49. Raines EW, Ross R (1982) J. Biol. Chem. *257*: 5154
50. Thomas KA, Rios-Candelore M, Fitzpatrick S (1984) Proc. Natl. Acad. Sci. USA *81*: 357
51. Rimsky L, Wakasugi H, Ferrara P, Robin P, Capdevielle J, Tursz T, Fradelizi D, Kutny RM (1986) J. Immunol. *136*: 3304
52. Nordin RP, Pumphrey JG, Rudikoff S (1987) J. Immunol. *139*: 813
53. Stanley ER, Guilbert J (1981) J. Immunological Methods *45*: 253
54. Wong GG, Witek JS, Temple PA, Wilkens KM, Leary AC, Luxenberg DP, Jones SS, Brown EL, Kay RM, Orr EC, Shoemaker C, Golde DW, Kaufman RJ, Hewick RM, Wang EA, Clark SC (1985) Science *228*: 810
55. Beutler B, Mahoney J, LeTrand N, Pekala P, Cerami A (1985) J. Exp. Med. *161*: 984
56. Tagaya Y, Okada M, Sugie K, Kasahara T, Kondo N, Hamuro J, Matsushima K, Dinarello CA, Yodoi J (1988) J. Immunol. *140*: 2614
57. Azhar S, Butte J, Reaven E (1987) Biochemistry *26*: 7047
58. Crankshaw DL, Crankshaw CL, Husby AD (1987) BioChromatogr. *2*: 196
59. Nandi J, Meng-Ai Z, Ray TK (1987) Biochemistry *26*: 4264
60. Walker LE, Reisfeld RA (1982) J. Immunological Methods *49*: R25
61. Kim BH, Rosenberry TL (1985) Biochemistry *24*: 3586
62. Bjorkman PJ, Strominger JL, Wiley DC (1985) J. Mol. Biol. *186*: 205
63. Bachmair A, Finley D, Varshavsky A (1986) Science *234*: 179
64. Lam KS, Neway JO, Gaucher GM (1988) Can. J. Microbiol. *34*: 30
65. Demain AL, Agathos SN (1986) Can. J. Microbiol. *32*: 208
66. Jensen SE, Westlake DWS (1983) Can. J. Microbiol. *29*: 1526
67. Hollander IJ, Shen YQ, Heim J, Demain AL, Wolfe S (1984) Science *224*: 610
68. Epstein CJ, Afinsen CB (1962) J. Biol. Chem. *237*: 2175
69. Acharya AS, Taniuchi H (1982) Mol. Cell. Biochem. *44*: 129
70. Price NC, Stevens F (1983) Biochem. J. *209*: 763

71. Fish WW, Danielsson A, Nordling K, Miller SH, Lam CF, Bjork I (1985) Biochemistry *24*: 1510
72. Brens DN, Plaisted SM, Havel HA, Kauffman EW, Stodola JD, Eaton LC, White RD (1985) Biochemistry *24*: 7662
73. Hager DA, Burgess RR (1980) Anal. Biochem. *109*: 76
74. Dame MC, Pierce EA, DeLuca HF (1985) Proc. Natl. Acad. Sci. USA *82*: 7825
75. Haggerty JG, Froehner SC (1981) J. Biol. Chem. *256*: 8294
76. Sakai D, Gorski J (1984) Biochemistry *23*: 3541
77. Bieger W, Scheele G (1980) Anal. Biochem. *109*: 222
78. Shaik MB, Guy AL, Pancholy SK (1980) Anal. Biochem. *103*: 140
79. Russell RR (1979) Anal. Biochem. *97*: 173
80. Tai PC, Zyk N, Citri N (1985) Anal. Biochem. *144*: 199
81. Brooks KP, Sander EG (1980) Anal. Biochem. *107*: 182
82. Ito K, Akiyama Y (1985) Biochem. Biophys. Res. Commun. *133*: 214
83. Benedek L, Dong S, Karger BL (1984) J. Chromatogr. *317*: 227
84. Katzenstein GE, Vrona SA, Wechsler RJ, Steadman BL, Lewis RV, Middaugh CR (1986) Proc. Natl. Acad. Sci. USA *83*: 4268
85. Fausnaugh JL, Kennedy LA, Regnier FE (1984) J. Chromatogr. *317*: 141
86. Wu SL, Benedek K, Karger BL (1986) J. Chromatogr. *359*: 3
87. Sharifi BG, Bascom CC, Khurana VK, Johnson TC (1985) J. Chromatogr. *324*: 173
88. Montelaro RC, West M, Issel CJ (1981) Anal. Biochem. *114*: 398
89. Swergold GD, Rubin CS (1983) Anal. Biochem. *131*: 295
90. Welling GW, Groen G, Welling-Wester S (1983) J. Chromatogr. *266*: 629
91. Bloemendal H, Groenewoud G (1981) Anal. Biochem. *117*: 327
92. Lu RC, Elzinga M (1977) Anal. Biochem. *77*: 243
93. Hjelmeland LM, Chrambach A (1984) Methods Enzymol. *104*: 305
94. van Renswoude J, Kempf C (1984) Methods Enzymol. *104*: 329
95. Klausner RD, van Renswoude J, Rivnay B (1984) Methods Enzymol. *104*: 340
96. Ashani Y, Catravas GN (1980) Anal. Biochem. *109*: 55
97. Furth AJ (1980) Anal. Biochem. *109*: 207
98. Amons R, Schrier PI (1981) Anal. Biochem. *116*: 439
99. Furth AJ, Bolton H, Potter J, Priddle JD (1984) Methods Enzymol. *104*: 318
100. Henderson LE, Oroszlan S, Koningsberg W (1979) Anal. Biochem. *93*: 153
101. Erickson JS, Paucker K (1979) Anal. Biochem. *98*: 214
102. Vecchio G, Righetti PG, Zanoni M, Artoni G, Gianazza E (1984) Anal. Biochem. *137*: 410
103. Riccio P, Rosenbusch JP, Quagliariella E (1984) FEBS Lettr. *177*: 236
104. Criado M, Aguilar JS, deRobertis E (1980) Anal. Biochem. *103*: 289
105. Kopaciewicz W, Regnier FE (1983b) Anal. Biochem. *133*: 251
106. Fish WW, Reynolds JA, Tanford C (1970) J. Biol. Chem. *245*: 5166
106a. Krueger JK, Kulke MH, Schutt C, Stock J (1989) BioPharm *2*: 40
107. Wolfe RA, Casey J, Familletti PC, Stein S (1984) J. Chromatogr. *296*: 277
108. Marston FA (1986) Biochem. J. *240*: 1
109. Dottin RP, Manrow RE, Fishel BR, Aukerman SL, Culleton JL (1979) Methods Enzymol. *68*: 513
110. Timmins JG, Petrovskis EA, Marchioli CC, Port LE (1985) Gene *39*: 89
111. Martinek K, Mozhaev VV, Berezin IV (1980) Biochim. Biophys. Acta *615*: 426
112. Saxena VP, Wetlaufer DB (1970) Biochemistry *9*: 5015
113. Swaisgood HE, Abraham P (1980) J. Dairy Sci. *63*: 1205
114. Pigiet VP, Schuster BJ (1986) Proc. Natl. Acad. Sci. USA *83*: 7643
115. Keesey J (1987) Biochemica information. A revised biochemical reference source. Boehringer Mannheim Biochemicals, Indianapolis, IN
116. Bell DJ, Hoare M, Dunnill P (1982) In: Fiechter A (ed) Advances in biochemical engineering/biotechnology. Springer Verlag, Berlin Heidelberg New York
117. Cordes A, Kula MR (1986) J. Chromatogr. *376*: 375
118. Lovrien R, Goldensoph C, Anderson PC, Odegaard B (1987) In: Burgess RR (ed) Protein purification. Alan R. Liss, New York
119. Chang SH, Gooding KM, Regnier FE (1976) J. Chromatogr. *125*: 103

120. Chang SH, Noel R, Regnier FE (1976) Anal. Chem. *48*: 1839
121. Regnier FE, Noel R (1976) J. Chromatogr. *14*: 316
122. Bidlingmeyer BA, Warren FV (1984) Anal. Chem. *56*: 1583A
123. Mikes W, Strop P, Zbrozek J, Coupek J (1976) J. Chromatogr. *119*: 339
124. Meyerson LR, Abraham KI (1986) Peptides *7*: 481
125. Yoshimoto T, Kita T, Ichinose M, Tsuru D (1982) J. Biochem. *92*: 275
126. Bennett HPJ, Browne CA, Solomon S (1983) Anal. Biochem. *128*: 121
127. Yau WW, Kirkland JJ, Bly DD, Stoklosa HJ (1978) J. Chromatogr. *149*: 465
128. Pfannkoch E, Lu KC, Regnier FE, Barth HG (1980) J. Chromatogr. Sci. *18*: 430
129. Kato Y, Hashimoto T (1983) J. High Res. Chrom. Comm. *6*: 45
130. Calam DH, Davidson J (1984) J. Chromatogr. *296*: 283
131. Josic D, Schutt W, Van Renswoude J, Reutter W (1986) J. Chromatogr. *353*: 13
132. Welling GW, Slopsema K, Welling-Wester S (1986) J. Chromatogr. *359*: 307
133. Corbett RJT, Roche RS (1984) Biochemistry *23*: 1888
134. Mahieu JP, Sebille B, Craescu CT, Rhoda MD, Beuzard Y (1985) J. Chromatogr. *327*: 313
134a. Technical Bulletin IL3, Whatman BioSystems, Inc, Clifton, NJ
135. Kornberg T, Gefter ML (1971) Proc. Natl. Acad. Sci. USA *68*: 761
136. Eley J (1969) Biochemistry *8*: 1502
137. Kopaciewicz W, Regnier FE (1983) Anal. Biochem. *128*: 196
138. Gooding KM, Schmuck MN (1984) J. Chromatogr. *296*: 321
139. Rounds MA, Regnier FE (1984) J. Chromatogr. *283*: 37
140. Gooding KM, Schmuck MN (1985) J. Chromatogr. *327*: 139
141. Kopaciewicz W, Rounds MA, Regnier FE (1985) J. Chromatogr. *318*: 157
141a. Technical Bulletin # 8700, JT Baker, Philipsburg, NJ
142. Lindblom H, Axio-Fredriksson AV, Cooper EH, Turner R (1983) J. Chromatogr. *273*: 107
143. Kato Y, Nakamura K, Yamazaki Y, Hashimoto T (1985) J. Chromatogr. *318*: 358
144. Radulovic LL, Kulkarni AP (1985) Biochem. Biophys. Res. Comm. *128*: 75
145. James S, Bennett HPJ (1985) J. Chromatogr. *326*: 329
146. Boyle DM, Wiehle RD, Shahabi NA, Wittliff JL (1985) J. Chromatogr. *327*: 369
147. Tomono T, Ikeda H, Tokunaga E (1983) J. Chromatogr. *266*: 39
148. Burchiel SW, Billman JR, Alber TR (1984) J. Immunological Methods *69*: 33
149. Danson MJ, Wood PA (1984) FEBS Lettr. *172*: 289
150. Siezen RJ, Kaplan ED, Anello RD (1985) Biochem. Biophys. Res. Comm. *127*: 153
151. FPLC Ion exchange and chromatofocusing. Principles and methods (1985) Laboratory Separation Division, Pharmacia Fine Chemicals, S-75182 Uppsala, Sweden
152. Soderberg L, Laas T, Low D (1982) Protides Biol. Fluids *29*: 955
153. Fagerstam LG, Lizana J, Axio-Fredriksson UB, Wahlstrom L (1983) J. Chromatogr. *266*: 523
154. Soderberg L, Muller RM, Fagerstam L (1983) Protides Biol. Fluids *30*: 661
155. Wahlstrom L, Muller RM, Fagerstam L, Stalberg R, Johansson JC, Fredriksson UB (1983) Protides Biol. Fluids *30*: 653
156. Wakefield LM, Cass AEG, Radda GK (1984) J. Biochem. Biophys. Methods *9*: 331
157. Tiselius A, Hjerten S, Levin O (1956) Arch. Biochem. Biophys. *65*: 132
158. Bernardi G (1973) Methods Enzymol. *27*: 471
159. Bernardi G, Kawasaki T (1968) Biochim. Biophys. Acta *160*: 301
160. Moss B, Rosenblum EN (1972) J. Biol. Chem. *247*: 5194
161. Brooks TL (1981) Hydroxylapatite. Fast flow and high resolution. Document # 8158-481. Calbiochem-Behring Corp, San Diego, CA
162. Matsushima K, Shiroo M, Kung H, Copeland T (1988) Biochemistry *27*: 3765
163. Steffensen R, Anderson JJ (1987) BioChromatogr. *2*: 85
164. Pearson JD, Regnier FE (1983) J. Liq. Chromatogr. *6*: 497
165. Hermodson M, Mahoney WC (1983) Methods Enzymol. *91*: 352
166. Bennett HPJ (1983) J. Chromatogr. *266*: 501
167. Cohen SA, Benedek K, Tapuhi Y, Ford JC, Karger BL (1985) Anal. Biochem. *144*: 275
168. Bansal SK, Love JH, Gurtoo HL (1985) Eur. J. Biochem. *146*: 23
169. Holm KA, Kupfer D (1985) J. Biol. Chem. *260*: 2027
170. Ingraham RH, Lau SYM, Taneja AK, Hodges RS (1985) J. Chromatogr. *327*: 77

171. Henderson LE, Sowder R, Copeland TD, Smythers G, Oroszlan S (1984) J. Virol. *52*: 492
172. Cohen SA, Schellenberg K, Benedek K, Karger BL (1984) Anal. Biochem. *140*: 223
173. Bietz JA, Cobb LA (1985) Cereal Chem. *62*: 332
174. Cohen SA, Benedek KP, Dong S, Tapuhi Y, Karger BL (1984) Anal. Chem. *56*: 217
175. Dolan JW, Snyder LR, Quarry MA (1987) Am. Lab. *19*: 43
176. Chang JP, Melander WR, Horvath C (1985) J. Chromatogr. *318*: 11
177. Bristow AF, Wilson C, Sutcliffe N (1983) J. Chromatogr. *270*: 285
178. Fett JW, Strydom DJ, Lobb RR, Alderman EM, Bethune JL, Riordan JF, Vallee BL (1985) Biochemistry *24*: 5480
179. Berkowitz SA, Henry MP, Nau DR, Crane LJ (1987) Am. Lab. *19*: 33
180. Padfield PJ, Case RM (1988) Anal. Biochem. *171*: 294
181. Alpert AJ (1986) J. Chromatogr. *359*: 85
182. Wetlaufer DB, Koenigbauer MR (1986) J. Chromatogr. *359*: 55
183. Weselake RJ, Jacobs HK (1983) Clin. Chim. Acta *134*: 357
184. Gooding DL, Schmuck MN, Gooding KM (1984) J. Chromatogr. *296*: 107
185. Lindahl L, Vogel HJ (1984) Anal. Biochem. *140*: 394
186. Vogel HJ, Lindahl L, Thulin E (1983) FEBS Lett. *157*: 241
187. Walsh PM, Valentine KA, Ngai PK, Carruthers CA, Hollenberg MD (1984) Biochemical J. *224*: 117
188. Szynkiewicz J, Stepkowski D, Brzeska H, Drabikowski W (1985) FEBS Lett. *181*: 281
189. Axen R, Porath J, Ernback S (1967) Nature *214*: 1302
190. Whiting PJ, Lindstrom JM (1986) Biochemistry *25*: 2082
191. Robb RJ, Kutry RM (1987) J. Immunol. *139*: 855
192. van Schravendijk MR, Biddison WE, Berger AE, Coligan JE, (1985) J. Immunol. *134*: 410
193. Zweig M, Schowalther SD, Bladen SV, Gilden RV, Arthur LO, Robey WG, Nara PL, Fischinger PJ (1988) AIDS Res. Human Retroviruses *4*: 51
194. Bouzerna N, Hammadi H, Richard C, Formstecher P, Dautrevaux M (1988) J. Immunological Methods *112*: 251
195. Hansen RS, Beavo JA (1982) Proc. Natl. Acad. Sci. USA *79*: 2788
196. Bailon P, Weber DV, Keeney RF, Fredericks JE, Smith C, Familletti C, Smart JE (1987) BioTechnol. *5*: 1195
197. Affinity chromatography. Principles and methods. (1979) Pharmacia Fine Chemicals, S-75104 Uppsala, Sweden
198. Anderson KK, Balny C, Douzou P, Bieth JG (1980) J. Chromatogr. *192*: 236
199. Khullar M, Scicli G, Carretero OA, Scicli AG (1986) Biochemistry *25*: 1851
200. Helseth DL, Veis A (1984) Proc. Natl. Acad. Sci. USA *81*: 3302
201. Sharf R, Weisman-Shomer P, Fry M (1988) Biochemistry *27*: 2990
202. Crute JJ, Wahl AF, Bambara RA (1986) Biochemistry *25*: 26
203. Lazarus GM, Henrich JP, Kelly WG, Schmitz S, Castora FJ (1987) Biochemistry *26*: 6195
204. Peumans WJ, Stinissen HM, Carlier AR (1982) Biochem. J. *203*: 239
205. Rutherford WM, Dick WE, Cavins JF, Dombrink-Kurtzman MA, Slodki ME (1986) Biochemistry *25*: 952
206. Day PM (1984) Eur. J. Biochem. *140*: 385
207. Madhok TC, Leung BS (1981) J. Steroid Biochem. *15*: 299
208. Kuhn RW, VestWeber C, Siiteri PK (1988) Biochemistry *27*: 2579
209. Gingrich JA, Amlaiky N, Senogles SE, Chang WK, McQuade RD, Berger JG, Caron MG (1988) Biochemistry *27*: 3907
210. Feldman SR, Pizzo SV (1986) Biochemistry *25*: 721
211. Furlong AM, Thomsen DR, Marotti KR, Post LE, Sharma SK (1988) Biotechnol. Applied Biochem. *10*: 454
212. Sulkowski E (1985) Trends Biotechnol. *3*: 1
213. Grussenmeyer T, Scheidtmann KH, Hutchinson MA, Eckhart W, Walter G (1985) Proc. Natl. Acad. Sci. USA *82*: 7952
214. Hatfield D, Thorgeirsson SS, Copeland TD, Oroszlan S, Bustin M (1988) Biochemistry *27*: 1179
215. Lew AM, Maloy WL, Coligan JE (1986) J. Immunol. *136*: 254
216. Walter G, Hutchinson MA, Hunter T, Eckhart W (1982) Proc. Natl. Acad. Sci. USA *79*: 4025

217. De Lisi C, Berzofsky J (1985) Proc. Natl. Acad. Sci. USA *82*: 7048
218. Karplus PA, Schulz GE (1985) Naturwissenschaften *72*: 212
219. Walter G (1986) J. Immunological Methods *88*: 149
220. Goudsdswaard J, van der Donk JA, Noordzij A, van Dam RH, Vaerman JD (1978) Scand. J. Immunol. *8*: 21
221. Fahnestock SR, Alexander P, Nagle J, Filpula D (1986) J. Bacteriol. *167*: 870
222. Underwood PA, Bean PA (1985) J. Immunological Methods *80*: 189
223. Nau DR (1986) Biochromatogr. *1*: 82
224. Ross AH, Herlyn D, Koprowski H (1987) J. Immunological Methods *102*: 227
225. Chen FM, Naeve GS, Epstein AL (1988) J. Chromatogr. *444*: 153
226. Tucker DF, Begent RHJ, Hogg NM (1978) J. Immunol. *121*: 1644
227. Stingl G, Wolff-Schreiner EC, Pichler WJ, Gschnait F, Knapp W, Wolff K (1977) Nature *268*: 245
228. Lillehoj EP, Myers NB, Lee DR, Hansen TH, Coligan JE (1985) J. Immunol. *135*: 1271
229. Rogers MJ, Siwarski DF, Shacter E, Maloy WL, Lillehoj EP, Coligan JE (1986) J. Immunol. *137*: 3006
230. Burgoyne RF, Bowles DK, Heckendorf A (1984) Am. Biotechnol. Lab. *2*: 1
231. Giulian GG, Moss RL, Graeser M (1983) Anal. Biochem. *129*: 227
232. Yudelson J (1983) Visualization of protein electrophorog rams by physical development. Kodak Laboratories Chemicals Bulletin # 74, Eastman Kodak Company, Rochester, NY
233. Weber K, Osborn M (1969) J. Biol. Chem. *224*: 4406
234. Laemmli UK (1970) Nature *227*: 680
235. Neville DM (1971) J. Biol. Chem. *246*: 6328
236. Wyckoff M, Rodbard D, Chrambach A (1977) Anal. Biochem. *78*: 459
237. Bothe D, Simonis M, von Dohren H (1985) Anal. Biochem. *151*: 49
238. Hjalmarsson SG, Baldesten A (1981) CRC Crit. Rev. Anal. Chem. *11*: 261
239. Arlinger L (1976) J. Chromatogr. *119*: 9
240. Holloway CJ, Battersby RV (1984) Methods Enzymol. *104*: 281
241. Vesterberg O (1968) Methods Enzymol. *22*: 559
242. Chapuis-Cellier C, Arnaud P (1981) Anal. Biochem. *113*: 325
243. Radola BJ (1984) Methods Enzymol. *104*: 256
244. Leary JJ, Brigati DJ, Ward DC (1983) Proc. Natl. Acad. Sci. USA *80*: 4045
245. Giulian GG, Moss RL, Greaser M (1984) Anal. Biochem. *142*: 421
246. Radola BJ (1973) Biochim. Biophys. Acta *295*: 412
247. Pious D, Krangel MS, Dixon LL, Parham P, Strominger JL (1982) Proc. Natl. Acad. Sci. USA *79*: 7832
248. Pober JS, Guild BC, Strominger JL (1978) Proc. Natl. Acad. Sci. USA *75*: 6002
249. Carr C, McCourt D, Cohen JB (1987) Biochemistry *26*: 7090.
250. O'Farrell PH (1975) J. Biol. Chem. *250*: 4007
251. Shackelford DA, Mann DL, van Rood JJ, Ferrara GB, Strominger JL (1981) Proc. Natl. Acad. Sci. USA *78*: 4566
252. Tracy RP, Katzmann JA, Kimlinger TK, Hurst GA, Young DS (1983) J. Immunological Methods *65*: 97
253. Wood DM, Dunbar BS (1981) J. Exp. Zool. *217*: 423
254. Towbin H, Gordon J (1984) J. Immunological Methods *72*: 313
255. Aebersold RH, Teplow DB, Hood LE, Kent SBH (1986) J. Biol. Chem. *261*: 4229
256. Dewald DB, Adams LD, Pearson JD (1986) Anal. Biochem. *154*: 502
257. Shively JE (1986) Methods of protein microcharacterization. Humana, Clifton, NJ
258. Hunkapillar MW, Lujan E, Ostander F, Hood LE (1983) Methods Enzymol. *91*: 227
259. Matsudaira P (1987) J. Biol. Chem. *262*: 10035
260. Andrews PC, Dixon JE (1987) Anal. Biochem. *161*: 524
261. Hewick RM, Hunkapillar MW, Hood LE, Dreyer WJ (1981) J. Biol. Chem. *256*: 7990
262. Schlabach TD, Wilson KJ (1987) J. Chromatogr. *385*: 65
263. Jorgenson JW, Rose DJ, Kennedy RT (1988) Am. Lab. *20*: 32
264. Allan JS, Coligan JE, Barin F, McLane MF, Sodroski JG, Rosen CA, Haseltine WA, Lee TH, Essex M (1985) Science *228*: 1091

265. Lillehoj EP, Salazar FHR, Mervis RA, Raum MG, Chan HW, Ahmad N, Venkatesan S (1988) J. Virol. *62*: 3053
266. Veronese FD, Copeland TD, DeVico AL, Rahman R, Oroszlan S, Gallo RC, Sarngadharan MG (1986) Science *231*: 1289
267. Rao RN (1984) Gene *31*: 247
268. Majumdar A, Rudikoff S, Adhya S (1987) J. Biol. Chem. *262*: 2326
269. White PJ, Millar G, Coggins JR (1988) Biochem. J. *251*: 313
270. Smith GE, Summers MD, Fraser MJ (1983) Mol. Cell. Biol. *3*: 2156
271. Pennock GD, Shoemaker C, Miller LK (1984) Mol. Cell. Biol. *4*: 399
272. Maiorella B, Inlow D, Shauger A, Harano D (1988) BioTechnol. *6*: 1406
273. Chakrabarti S, Robert-Guroff M, Wong-Staal F, Gallo RC, Moss B (1986) Nature *320*: 535
274. Moss B (1985) Immunol. Today *6*: 243
275. Strauss M, Kiessling U, Platzer M (1986) Biol. Zentralbl. *105*: 209
276. Temin HM (1986) Retrovirus vectors for gene transfer: Efficient integration into and expression of exogeneous DNA in vertebrate cell genomes. In: Kucherlapati R (ed) Gene transfer. Plenum, New York, p 149
277. Bolstein D, Shortle D (1985) Science *229*: 1193
278. Bott R, Ultsch M, Kossiakoff A, Graycar T, Katz B, Power S (1988) J. Biol. Chem. *263*: 7895
279. Pantoliano MW, Whitlow M, Wood JF, Rollence ML, Finzel BC, Gilliland GL, Poulos TL, Bryan PN (1988) Biochemistry *27*: 8311
280. Young J, Davis J (1983) Proc. Natl. Acad. Sci. USA *80*: 1194
281. Huynh TV, Young RA, Davis RW (1984) In: Glover D (ed) DNA cloning: A practical approach. Vol 3. IRL, Oxford, UK
282. Carroll SB, Laughon A (1987) In: Glover D (ed) DNA cloning: A practical approach. Vol. 3. IRL, Oxford, UK
283. Germino J, Bastia D (1984) Proc. Natl. Acad. Sci. USA *81*: 4692
284. Nagai K, Thogerson HC (1987) Methods Enzymol. *153*: 461
285. Schoner RG, Ellis LF, Schoner BE (1985) BioTechnol. *3*: 151
286. Sung WL, Yao FL, Zahab DM, Narang SA (1986) Proc. Natl. Acad. Sci. USA *83*: 561
287. Nilsson B, Holmgren E, Josephson S, Gatenbeck S, Philipson L, Uhlen M (1985) Nucleic Acids Res. *13*: 1151
288. Ikehara M, Ohtsuka E, Tokunaga T, Nishikawa S, Uesugi S, Tanaka T, Aoyama Y, Kikyodani S, Fujimoto K, Yanase K, Fuchimura K, Morioka H (1986) Biochemistry *83*: 4695
289. Moks T, Abrahmsen L, Osterlof B, Josephson S, Ostling M, Enfors S, Persson I, Nilsson B, Uhlen M (1987) BioTechnol. *5*: 379
290. Sassenfeld HM, Brewer SJ (1984) BioTechnol. *2*: 76
291. Hopp TP, Prickett KS, Price VL, Libby RT, March CJ, Cerretti DP, Urdal DL, Conlon PJ (1988) BioTechnol. *6*: 1204
292. Lin Y, Robb RJ, Gray JE, Simon P (1988) J. Immunol. *141*: 3847
293. McCormick D (1988) BioTechnol. *6*: 158
294. Cazes J, Nunogaki K (1987) Am. Lab. *19*: 126
295. Compton SW, Brownlee RG (1988) BioTechniques *6*: 432

Lipid Biosynthesis in Oleaginous Yeasts

G. Turcotte* and N. Kosaric
Department of Chemical and Biochemical Engineering, The University of Western Ontario, London, Ontario, Canada, N6A 5B9

1 Introduction ... 74
2 Effects on Total Intracellular Microbial Lipid Synthesis ... 77
2.1 Nitrogen Source ... 78
2.2 Carbon Source ... 78
2.3 Vitamins ... 79
2.4 Temperature ... 79
2.5 pH ... 79
2.6 Aeration ... 80
2.7 Redox Potential ... 82
2.8 Temporal Pattern ... 82
2.9 Growth Rate ... 82
2.10 Combined Effects of the Above Parameters ... 84
3 Effects on Individual Fatty Components of the Intracellular Microbial Lipids ... 85
3.1 Growth Rate ... 85
3.2 Temperature ... 85
3.3 Temporal Pattern ... 85
4 Effects on the Fatty Acid Composition of Intracellular Microbial Lipids ... 86
5 Extracellular Microbial Lipids ... 88
6 Concluding Remarks ... 89
7 Nomenclature ... 90
8 References ... 90

With the increasing demands for fats and oils for edible and industrial purposes, many assessments have been made to find new possible lipid sources other than from the conventional plants and animals. Lipids from microbial origin could represent such an alternative since microorganisms can store up to 65% fat in the cells and can be harvested in a minimum time, unaffected by climatic changes. However, unless the microbial oil, or one of its constituents, represent a high value specialty product, its production would be uneconomical.

This paper reviews the environmental conditions affecting the composition of the microbial oil (lipid classes and fatty-acyl distribution) from oleaginous yeasts. This approach will show that it is possible to "tailor-make" an oil with special properties, albeit in a restricted way.

* Present address: Department of Chemical Engineering, University of Ottawa, Ottawa, Ontario, Canada, K1N 6N5

1 Introduction

Demand for fats and oils is ever-increasing, whether it is for edible foods or for non-edible purposes. Interest in fat-producing microorganisms as a substitute for natural (animal and vegetable) fats and oils resulted in numerous investigations, mainly because microbial lipids and natural fats and oils share some important characteristics:

a) their fatty acids have approximately the same distribution and show similar variations in their relative proportions (see Tables 1 and 2);
b) the predominant type of acids in microbial and natural triglycerides are usually 1-oleyl-(or palmityl)-2-oleyl-3-palmityl-(or oleyl)-glycerol; i.e., SUU, UUS, or SUS type (where S is any saturated acid and U is any unsaturated acid) [1]; and
c) both natural and microbial lipids contain the same types of soluble, minor constituents such as pigments (carotenoid, chlorophyl, etc.), sterols (phytosterols in plant fats, cholesterol in animal fats, ergosterol in yeast), phospholipids (lecithin), lipoproteins, glycolipids, hydrocarbons, vitamin E (tocopherols), vitamin A (from carotenes), vitamin D (calcipherol), waxes, esters of long-chain alcohols and fatty acids, ethers and degradation products of fatty acids, proteins, and carbohydrates. Most of these compounds are removed by processing and some are valuable byproducts (see Table 3).

Roughly two-thirds of the world's production of fats and oils is utilized for human food. Non-edible uses depend mainly on the properties of the various fatty acids in the fats. Some sulfurized triglycerides which contain long-chain fatty acids and are monounsaturated are being used as specialized lubricant additives (e.g. hog fat, wax ester jojoba oil [14]). Derivatives of fatty acids (see Table 4) are used in surface coatings, plastics, detergent, lubricants, etc., where the long hydrocarbon chains confer needed plasticity, surface activity, or lubricity [15–17].

Microorganisms can not only produce lipids in their cells, but they can also excrete related products which have special properties. Examples are: surface-active agents [18,19], hydroxy fatty acids [20], saturated and unsaturated wax esters [21] and lipids with odd-number fatty acids.

Thus, using a microbial route, many different products are potentially available. Comparing Tables 1, 3, and 4, one can see that where a specific application and very desirable property is required, the price of the commodity is likely to be high. Such is the case with olive oil for which no substitute is accepted (high oleic acid content), cocoa butter for its peculiar melting behaviour (due to its specific fatty acid composition) and the undecylenic acid and octyl alcohol for use in the cosmetics industry. However, not all products reported in these tables can be efficiently replaced by microbial lipids. The rate and amount of the lipid compound produced, its yield and purification steps required are but a few examples of limitations encountered.

A product which can be potentially available in a large quantity from microbial cells is the intracellular lipids which can account for up to 65% of the dry biomass of oleaginous yeasts [22–26]. Oleaginous yeasts are recognized as having a high potential for fat production and a few patents were issued concerning conditions for their cultivation [13–27]. They would be relatively well accepted as a food component. Not only should the yeasts produce large amounts of lipids, but these must be of the appropriate form and possess a specific desirable fatty acid distribution. It will be

Table 1. Oil content, fatty-acyl composition, and prices of some natural fats and oils

Fats or Oils		Sesame	Cocoa butter[a]	Lard	Corn	Soybean	Sun-flower	Olive	Linseed	Tung[c]	Palm
Refs.		2)	3)	4)	2)	4)	4)	2)	2)	5)	2)
1. Oil content, % wt/wt		50–55		100	50 germ	22–32	36–40	30–60 fruit	32–40	51–67	
2. Fatty-acyl composition % wt/wt	12:0	—	—	0.5	—	—	—	—	—	—	—
	14:0	—	0.7	3.5	—	0.5	0.5	tr	—	—	1
	16:0	10	25.2	28.5	13	11	6.5	14	6	8	48
	16:1	—	—	4	—	—	—	2	—	—	—
	18:0	5	35.5	20	4	4.5	3.5	2	4	—	4
	18:1	40	35.2	39	29	22	23	64	22	15	38
	18:2	45	3.2	3	54	53	64.5	16	16	75–78	9
	18:3	—	0.2	0.5	—	8	0.5	—	52	—	—
	20:0	—	—	0.5	tr	0.5	1	tr	tr	—	—
	>20:1	—	—	—	tr	0.5	0.5	—	—	—	—
3. Applications		Edible	Edible cosmetics	Edible	Edible	Edible	Edible	Edible	Paints and varnishes		Edible
4. Price[b], US-$/kg		2.20	4.71	0.28	0.44	0.31	0.34	2.31	0.55	0.68	0.35

[a] Fatty-acyl composition in mole %;
[b] from Chemical Marketing Reporter and the Journal of Commerce, Nov. 1986;
[c] saturated 8% wt/wt

Table 2. Fat content and fatty-acyl composition of some microorganisms

Microorganism		A	B	C	D	E	F	G	H	I	J
Refs.		6)	7)	8)	9)	10)	11)	12)	12)	12)	13)
1. Oil content, % wt/wt		[a]	64	[a]	62	40	12	42	12.5	5	62.5
2. Fatty-acyl											
composition	16:0	13	20	32	12	18.1	13.3	37	19.8	17	30.5
% wt/wt	16:1	1	2	—	1.5	—	0.8	1	2.0	18.1	—
	18:0	5	1	3	2	16.9	5.1	14	2.4	4.8	10.7
	18:1	14	42	37	82.5	58.5	1.2	36	48.6	41.8	54.1
	18:2	61	21	10	2	3.5	77	8	26.3	16.6	—
	18:3	7	8	5	—	2.0	—	—	—	—	—

A *Penicillium chrysogenum*
B *Rhodotorula gracilis*
C *Rhodotorula graminis*
D *Rhodotorula glutinis*
E *Rhodosporidium toruloides*
F *Saccharomyces cerevisiae*
G *Candida* 107 (now *Candida* NCYC 911)
H *Hansenula anomala*
I *Candida lipolytica*
J *Lipomyces starkeyi*
[a] not available

shown later how these parameters can be affected by environmental conditions during cultivation such as temperature, pH, nutrients, oxygen availability, etc. There is still insufficient information available as to which factors are responsible for a change of lipid composition, and particularly whether and to what extent synergistic effects play a role in microbial lipid synthesis.

The following discussion focusses on conditions of incubation affecting 1) the total amount of lipids produced in the cell, 2) the types of lipids synthesized (triglycerides, phospholipids, sterols, etc.), 3) the fatty acid distribution, and 4) the extracellular lipid production.

Table 3. Prices and applications of some minor constituents of natural fats

Product	Price[a] US-$ per kg	Applications
Lecithin	0.79	Emulsifier in foods and feeds, in cosmetics, soaps, leather, textiles, paints, lubricants, and pharmaceuticals
Glycolipids	40.	Emulsifiers or de-emulsifiers
Vitamin E (tocopherol)	50.	Food supplement
Vitamin A	33.	Food supplement
Carotene	59.	Pro-Vitamin A
Vitamin D (cholecalciferol)	24.	Food supplement

[a] from Chemical Marketing Reporter, Nov. 1986

Table 4. Prices and applications of some minor individual fatty acids and derivatives

Product		Price US-$ per kg	Applications
Caprylic acid	$CH_3(CH_2)_6COOH$	1.61	Intermediate in manufacture of esters for perfumes
Capric acid	$CH_3(CH_2)_8COOH$	1.32	Artificial fruit flavors and perfumes
Lauric acid	$CH_3(CH_2)_{10}COOH$	1.43	Soaps; production of docecylamine → germicidal quaternary NH_4 salt
Myristic acid	$CH_3(CH_2)_{12}COOH$	2.86	Production of lauric acid; shaving creams, shampoos and cosmetics
Oleic acid	$CH_3(CH_2)_7CH{=}CH(CH_2)_7COOH$	1.01	Liquid soaps, disinfectants, shampoos, textile lubricants, carbon paper, stencil and printing inks; rayon finishes; waxes and polishes; ore flotation; latex dispersant
Palmitic acid	$CH_3(CH_2)_{14}COOH$	1.17	Paints, varnishes, lacquers
Undecylenic acid	$CH_2{-}CH(CH_2)_8COOH$	5.94	Cosmetics, Nylon II, heptaldehyde
Ricinoleic acid	$CH_3(CH_2)_5CH(OH)CH_2CH{=}CH(CH_2)_7COOH$	1.74	Resins (drying oils), lubricating greases and oils, disinfectants, hair products
Stearic acid	$CH_3(CH_2)_{16}COOH$	0.57	Cosmetics, creams, shaving soaps, pharmaceuticals
Octyl-alcohol	$CH_3(CH_2)_6CH_2OH$	3.08	Perfume and cosmetic additive; plasticizer, frother, polymerization; anti-bacterial, anti-fungal properties
Myristil alcohol	$CH_3(CH_2)_{12}CH_2OH$	—	Emulsifier in cosmetic creams
Cetyl alcohol	$CH_3(CH_2)_{14}CH_2OH$	1.50	Extensively used in cosmetic industry; brushless shaves; vanishing creams
Stearyl alcohol	$CH_3(CH_2)_{16}CH_2OH$	0.53	Cosmetics
Others			
(a) Emulsifiers (food)			mono-, di-, triglycerides; polyglycerol esters; propylene glycol mono-saturated; sorbitan esters; stearoyl lactylates
(b) Surfactants			fatty alcohol sulfates, C_8-C_{10}: tauxate (Igepon T), $C_{18:1}$: alkylolamides; Na laureylsarcosinate
(c) Plasticizers			aliphatic esters (> C_4) of dibasic acids (phthalic, etc.) expoxidized unsaturated natural esters (soybean oils)
(d) Waxes			polyethylene glycols; ester waxes; ketones; amines, C_6-C_{18}

2 Effects on Total Intracellular Microbial Lipid Synthesis

Although certain yeasts can be induced to produce as much as 65% wt/wt fat in cells [1,13,22,23,26], values around 35% are more common [28-31].

Microbial lipids are synthesized throughout growth as a membrane component and as a carbon storage reserve but the accumulation of fat in most organisms is said not to begin until a nutrient, other than the carbon source, has become exhausted from the medium. Nitrogen was found to be the limiting nutrient for yeasts [31-38]. However, the yeast *Cryptococcus albidus* var. *albidus* was shown by Hansson and Dostálek [39,40] to produce around 25% lipids in a nitrogen-rich medium whereas the lipid content reached approximately 32% when the nitrogen of the medium was limiting.

Ratledge and Hall [30] found that a dual limitation of nitrogen and phosphate increased neither the amount of lipids produced nor the lipid yield (g lipid produced per g glucose metabolized). This dual limitation of nitrogen and phosphate on *Rh. glutinis* contradicts the previous results of Enebo et al. [22] on *Rhodotorula* yeast, and those of Gill et al. [29] on *Candida* 107 (now *Candida* NCYC 911). Depletion of phosphate, sulfate and iron have been shown to induce fat formation in *Rh. gracilis* [41, 42] but not all organisms will accumulate fat if a nutrient other than the nitrogenous one becomes exhausted. This further suggests that an optimum balance of nutrients for maximum fat formation is characteristic of the lipogenic microbes.

2.1 Nitrogen Source

The effect of the nitrogen sources NH_4Cl, asparagine and glutamate were studied by Evans and Ratledge [43] on 17 species and strains of yeasts. They found that only a small number of yeasts showed any variation in lipid content, the most significant of these being strains of *Rhodosporidium toruloides*. Similar results were found by Hansson and Dostálek [40] with *Cryptococcus albidus* var. *albidus* who specified that the growth curves were almost identical regardless of the nitrogen source used. *Rs. toruloides* CBS 14 accumulated fat in larger amounts (45 to 53%) when it was grown in the presence of urea, threonine, aspartate, asparagine, glutamate and arginine as opposed to around 20% when grown in the presence of inorganic nitrogen sources (e.g., NH_4Cl and $NaNO_3$) or with mixtures of the organic and inorganic types [44]. They hypothesized that the rate of lipid production is influenced by the products of catabolism (namely enzymes) of the medium nitrogen source rather than being the result of direct stimulation by the nitrogen compound itself. Furthermore, they argued that the concentration of intracellular NH_4^+ is probably of considerable regulatory significance during the inital stages of lipid biosynthesis.

2.2 Carbon Source

Production of one ton of lipid requires about five tons of substrate [24] and this cost is the most important economic element in the process. Availability of a cheap carbon source will make production of a wider range of edible oil substitutes possible than with higher-priced carbon sources which restrict the possibilities to higher-priced specialty lipids such as cocoa butter substitutes. A few examples of waste materials used as carbon sources for investigation of lipid production are presented in Table 5. Since each individual waste material must be assessed as to the composition and concentration of its components and many microorganisms must be screened in as many experimental conditions, the amount of work required to find the right combination of microbe-waste material-experimental conditions is very large. Furthermore, acclimatization of the microbes to their environment might result in better growth and lipid synthesis.

Table 5. Examples of waste materials for lipid production

Waste material	Yeast culture	Refs.
Bananas and other food wastes	*Candida curvata* D	[45]
Beet pulp extract	*Rhodotorula gracilis*	[46]
Corn steep liquor	*Rhodosporidium toruloides*	[47]
Molasses	*Rh. gracilis*	[28, 32, 46, 48]
Potato chip waste water	*Rh. glutinis*	[49]
Starch	*C. utilis*	[50]
	Saccharomycopsis fibuliger/Rs. toruloides	[51]
Vinasse of wine distilleries	Yeasts	[52]
Whey	*Rh. gracilis*	[28, 48]
Whey permeate	*C. curvata* D	[53]
	C. curvata or *Candida* NCYC 911	[54]

2.3 Vitamins

A detailed vitamin requirement study by Allen et al. [32] demonstrated that *Rh. gracilis* required only thiamine and pantothenic acid for maximum growth, but evidence was obtained that no vitamins or growth factors were required during the fattening phase, confirming a previous observation of Steinberg and Ordal [55].

2.4 Temperature

The effect of temperature on the lipid synthesis of yeast generally follows that of most common organisms: the lipid content rises as the environmental temperature is dropped below that for optimal growth. But in studying *Candida* yeasts, Kates and Baxter [56], and Thorpe and Ratledge [57] agreed that an increase in temperature from 10 °C to 25 °C, and from 30 °C to 40 °C, respectively, resulted in an approximately constant amount of total lipids although changes in individual lipids and fatty acid composition occurred. The lipid content of *Rh. gracilis* was said to decrease with an increase of temperature from 27 °C to 35 °C [58]. Goulet [34] compared the energy of activation for fat synthesis by *Rh. glutinis* Mac Y-3748 with that for its growth, and concluded that temperature (range of 23 °C to 31 °C) had a lesser effect on the rate of lipid synthesis than on growth.

2.5 pH

Although Enebo et al. [22] suggested that pH should be controlled during growth to minimize the effect of too sudden pH changes, no consistent changes in the lipid production due to pH are reported in the literature. This seems to indicate that pH may play only a secondary role. Goulet [34] found that the fat content increased linearly from 20 to 22% with an increase in pH, controlled between 3.0 and 8.0, whereas Kessell [7] working at pH 3.0 to 6.0 found that growth of the organism was retarded at low pH values but that the lipid content decreased from 63 to 40%, although the final concentration of lipids extracted remained the same.

The effect of initial pH of the medium on the lipid composition of *Candida stellatoidea* was studied by Guarneri et al. [59]. The initial pHs chosen were 4.0, 6.5, 8.0, and 10.0, and the final pHs after culturing 48 h were 6.93, 8.29, 7.80, and 8.95, respectively. Since the temporal pattern on pH variation was not given, one can only assume that the cells reacted quite differently to changes in initial environmental pH conditions; thus, the change in lipid production reported cannot be attributed directly to the effect of pH alone, but to a combination of the pH effect and the change in the cell physiology as a function of time and initial conditions.

2.6 Aeration

Oxygen has a pronounced effect on the growth and general metabolism of any organism. Although aeration was recognized as an important factor in the unsaturation of the fatty acids [60–63], this parameter has been rarely quantified. Pan et al. [64] found that varying the rate of aeration from 0.1 to 1 vvm[1] had no effect on the yield of the cells of *Rh. gracilis* or its fat content although it slightly influenced the rate of sugar utilization. Also, aeration in shake flasks (80 ml of medium in a 500 ml flask) was found to be nearly equivalent to an air sparging ratio of 0.333. On the basis of these results, they concluded that from a practical standpoint, aeration ratios higher than 0.333 were not advantageous.

An empirical aeration ratio of 1.0 has been used by many investigators [7,32,55]. Spotholz et al. [65] established that supplying 3.0 g dissolved oxygen per liter of medium per hour to *Rh. gracilis* satisfied the requirements for the yeast to grow exponentially until the substrate became limiting. Kessell [7] determined the critical dissolved oxygen concentration at 325 $\mu g\ l^{-1}$ during exponential growth of the same micro-organism and observed that the dry weight of the cells could be raised from 9.5 to 15 $g\ l^{-1}$ at the end of the exponential phase by raising the rate of agitation from 600 to 1200 rpm, thus increasing the oxygen dissolution rate. Combs et al. [66] reported the same trend with *Candida albicans* but low oxygen pressures (2 and 140 mm Hg) did not produce significant changes in the amounts of total lipids and phospholipid fatty acids of *Rh. gracilis*, while the level of non-phospholipid fatty acids was slightly depressed [67]. When fat accumulation was allowed to occur (by using N-limiting medium) in continuous culture of *Candida* NCYC 911 and *Rh. gracilis*, Ratledge and Hall [68] found that oxygen demand dropped dramatically. This had been anticipated since fat is essentially a chemically reduced material which requires no oxygen for its synthesis from malonyl-coenzyme A (CoA) units. However, Turcotte [69] showed (see Fig. 1) dissolved oxygen (DO) concentration profiles during batch culture of *Rhodosporidium toruloides* ATCC 10788 which seemed to indicate that oxygen might be required shortly after nitrogen depletion when the cells shift their metabolism toward lipid accumulation. The oxygen demand dropped afterwards. Choi et al. [10] increased the dissolved oxygen in continuous cultures of *Rh. gracilis* from 45 to 234 μM which resulted in an increase in lipid content while the degree of unsaturation remained relatively constant.

It must be recognized, however, that all of the above findings can only serve as guidelines; the many interactions between dissolved oxygen concentration and

[1] vvm = volume of air per volume of medium per minute

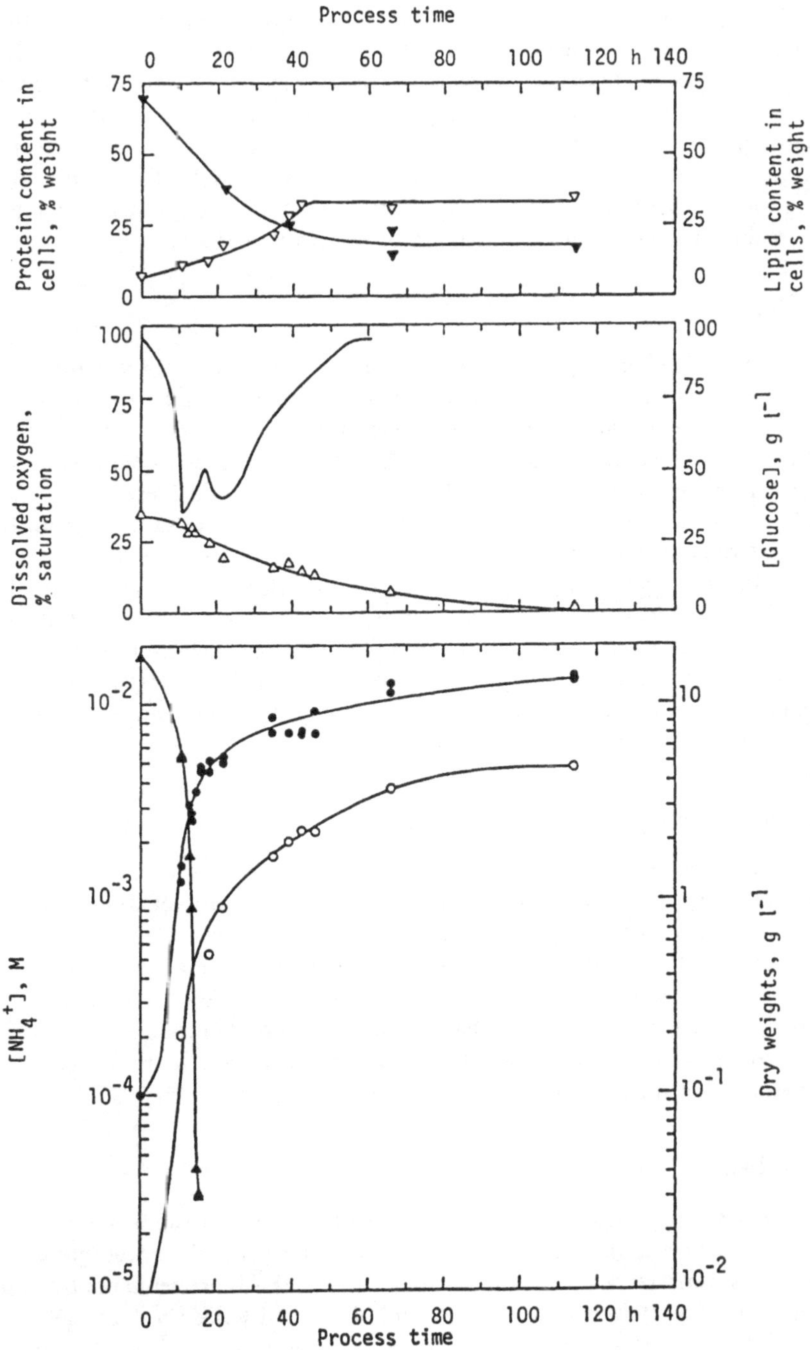

Fig. 1. Growth and lipid production by *Rs. toruloides* at 30 °C and pH 6, in a controlled process (Chemap fermenter; 300 rpm; aeration rate = 0.6 vvm; 15 liters of medium; mechanical antifoam control). ▼ Protein content; ▽ lipid content; — [dissolved oxygen], in medium; △ [glucose]; ▲ [N] as NH_4^+, in the supernatant; ● [Cells] (d.w.); ○ [intracellular lipids] (d.w.)

experimental temperature, medium composition and oxygen requirements of the organisms have not been investigated. It will become possible to study these interactions after some improvements are made on the measurement and control of dissolved oxygen in a bioreactor, namely, a) the quick quantification of oxygen concentration in absolute units (i.e., mg ml^{-1}) in a medium where the components and their concentrations are varying, and b) the control of oxygen at a constant level without changing the shearing environment (e.g. introducing a mixture of pure oxygen and air as opposed to manipulating the agitation rate and/or the gas flow rate).

2.7 Redox Potential

The redox potential has been widely adopted as an index to the quality control of anaerobic processes. In aerobic cultures, electrode potential is said to yield more information about the oxidative status than the concentration of dissolved oxygen [70-72]. Although Shibai et al. [73] and Akashi et al. [74] found the redox potential to be an important parameter in the production of inosine and amino acids, respectively, no studies could be found in the literature relating redox potential to fat formation. It must be emphasized that the redox potential indicates the oxidative status of aerobic cultivations at low dissolved oxygen (DO) concentrations (less than 1% saturation), and that when the DO becomes limiting, the energy production rates within the cell are reduced. The latter is in contradiction to the production of lipids which is an energy-requiring process. Wimpenny and Necklen [75] found that ATP associated enzymes reached a maximum value corresponding to about 300 mV which coincided with high dissolved oxygen values and maximum cell yields.

2.8 Temporal Pattern

Age of the culture definitely plays an important role although only a few investigators have researched this field. Steinberg and Ordal [55] found that the ratio of fat to non-fat of *Rh. gracilis* increased linearly with time, but no maximum was reported. Allen et al. [32] reported that the yield of fat from sugar was at a maximum at approximately 14 to 23 h, then decreased rapidly. However, the amount of fat could be increased by adding sugar daily during the fattening phase. *C. utilis* seemed to produce its highest amount of lipid in the stationary phase (after 60 h growth) in a batch system [76], whereas the amount of lipids from *C. albicans* decreased with time [66].

2.9 Growth Rate

In closed systems, such as batch cultures, the composition of the medium is modified by each successive generation of cells and ultimately results in unbalanced growth. Steady states, however, are said to be obtained when the yeast is grown in controlled open systems; e.g. in a chemostat. Goulet [34] and Yoon and Rhee [38,77] and Ghose et al. [78] found that decreasing the dilution rates to around 0.02 h^{-1} resulted in a percentage of lipids in the biomass similar to that of batch cultures. At this extremely low dilution rate, the cells spend an average residence time of 50 h which might be at least 10 h longer than the time required for depletion of nitrogen (based on the batch results of Yoon and Rhee [38]). Thus, the relationship "dilution rate equals growth

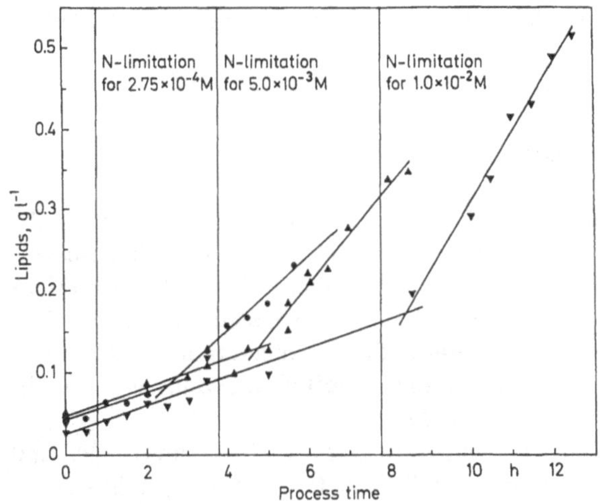

Fig. 2. Lipid production by *Rs. toruloides* from different initial levels of nitrogen at 30 °C, in shake flasks. Initial $[NH_4^+]$: ● 2.75×10^{-4} M; ▲ 5.0×10^{-3} M; ▼ 1.0×10^{-2} M; — linear regression equation

rate" becomes nonsensical since for both cells and lipids the growth rates prior to nitrogen depletion are not the same as rates due to fattening of the cells after nitrogen exhaustion.

The results in continuous systems could probably be interpreted in the following ways: a) only a restricted range of dilution rates allows sufficient time after nitrogen-depletion for lipid production to occur, whereas lower or faster dilution rates produce too slow or too rapid growth rates for the cells to ever reach nitrogen exhaustion, or b) the rate of nitrogen supplied to the cells is responsible for the lipid production rates. No information was found in the literature to favor either one or the other argument. Turcotte [69]) tested the second hypothesis by varying the concentration of initial nitrogen as $(NH_4)_2SO_4$ alone to one gram of "lean" cells (<7% fat content) of *Rs. toruloides* ATCC 10788 in batch cultures. The accumulation of lipids always started sometime after all the nitrogen was exhausted from the medium (see Fig. 2) although this time was shorter for higher initial concentration of nitrogen. It must be pointed out that around 86% of the glucose concentration remained after 12 h, ruling out the possibility of glucose limitation.

In other conditions (higher fat content of the inoculum and 2% nitrogen present as yeast extract), the accumulation of lipids could start up to 4 h prior to NH_4-limitation.

Although the results seem to indicate that a certain high level of nitrogen could rapidly produce more lipids, the specific lipid productivity (per g of lipid-free cellular material) was shown to be statistically equivalent. Furthermore, Fig. 2 shows that the production of the growth associated lipids (prior to NH_4^+-limitation) was independent of the range of initial concentration of nitrogen (all slopes of lipid concentration vs time of incubation were found to be statistically equivalent).

Thus, regardless of the amount of nitrogen supplied to the cells, the specific growth rate for lipids produced intracellularly seems to be a constant, provided no other nutrient becomes limiting.

2.10 Combined Effects of the Above Parameters

It is a common practice to assess the effect of the above parameters in a one-variable-at-a-time approach, where one variable is fixed at an optimum and another variable is studied. Once the optimum value of the second parameter is found, the conditions are fixed at the two optima and the third variable is investigated, and so on. This approach is tedious and at the same time assumes that no interaction exists between the said parameters. None of the literature studies reported above investigated the combined effect of any of the parameters involved.

We investigated the combined effect of temperature (25 °C, 30 °C and 35 °C) and pH (4.5, 6 and 7.5) and the results are presented in Table 6. It was found that the specific growth rates of both cells and intracellular lipids before nitrogen-depletion did not vary significantly from one set of conditions to the other and neither did the specific growth rate of the lipids accumulating after nitrogen depletion. However, a trend was present in the values of the maximum lipid content of the cells: the lipid content at lower pH was always 4 to 5% more than that at the higher pH and this regardless of the temperature level. Alternately, lowering the temperature increased the lipid content of the higher temperature level by around 10% more. This occurred at all pH's investigated. It was shown by statistical analysis that the effect of pH was not significant while that of temperature was. This helped prove that pH plays a secondary role in lipid production, but that its effect, combined with that of temperature, is important.

Table 6. Results from a three-level factoral design with *Rs. toruloides*

x_1	x_2	y_1	y_2	y_3	y_4
−1	−1	34.5	0.197	0.139	0.0230
0	−1	30.8	0.269	0.259	0.0307
1	−1	26.8	0.304	0.383	0.0214
−1	0	39.7	0.152	0.241	0.0225
0	0	34.6	0.283	0.404	0.0244
1	0	30.0	0.308	0.319	0.0612
−1	1	40.5	0.745	0.710	0.0858
0	1	35.7	0.297	0.298	0.0239
1	1	30.2	0.274	0.277	0.0150

x_1 = Coded temperature values
−1 = 25 °C
0 = 30 °C
1 = 35 °C

x_2 = Coded pH values
1 = 4.5
0 = 6
1 = 7.5

y's = Average responses at the test conditions
y_1 = Max. intracellular lipid content, % dry weight
y_2 = μ cells (before N-depletion), $gl^{-1}h^{-1}$
y_3 = μ lipids (before N-depletion), $gl^{-1}h^{-1}$
y_4 = μ lipids (after N-depletion), $gl^{-1}h^{-1}$

3 Effects on Individual Fatty Components of the Intracellular Microbial Lipids

The intracellular microbial lipids usually consist of mono-, di-, triglycerides, phospholipids, sterols, sterol esters, and free fatty acids. Occasionally, squalene and ergosterol are also reported. These are part of the primary products of metabolism [4)]. Triglycerides are found to constitute the major portion of the fat of mature cells grown in nitrogen-limiting conditions in many *Candida* species [12,26,29)], certain strains of *Lipomyces* [37)], and *Rhodotorula* (67% of total lipid [38,77)]), whereas phospholipids predominate in *Saccharomyces cerevisiae* [79)], a non-fat producer. Pantothenic acid deficiency was shown by Furukawa and Kimura [80)] to shift the formation of triglyceride, sterol, diglyceride, and sterol esters into the production of free fatty acids. The total amount of phospholipids was not compared at that time so that it is not possible to speculate on the crucial partition of phosphatidic acid between the two branchings (towards triacylglycerol synthesis or that of phosphoglycerides) in the regulation of lipid synthesis [81,82)].

3.1 Growth Rate

When Yoon and Rhee [38,77)] studied *Rh. glutinis* in continuous culture, they found that the total amount of lipids decreased when dilution rates increased from 0.02 h^{-1} to 0.30 h^{-1}, the percentage weight of the neutral lipid fraction decreased from 83.6 to 54.7, whereas glycolipid and phospholipid fractions increased from 5.7 to 21.2, and from 10.7 to 24.2, respectively. These results in nitrogen-limiting conditions were of an opposite trend to those in carbon-limited cultures.

3.2 Temperature

Hunter and Rose [83)] found that lowering the growth temperature of batch cultures was accompanied by a increase in the contents of total fatty acids, triacylglycerols and phospholipids, but had little effect on the contents of diacylglycerols, free fatty acids, sterols, or sterol esters.

3.3 Temporal Pattern

Figure 3 shows the profile of the main classes of lipids while incubating *Rs. toruloides* at 35 °C and pH 7.5 [69)]. Although lipids started to accumulate after 17 h, their composition had already levelled off. The peculiar patterns of phospholipids, sterols and glycolipids cannot be explained as yet and might be particular to the conditions chosen. Nevertheless, these results seem to contradict those of Hunter and Rose [83)] presented above.

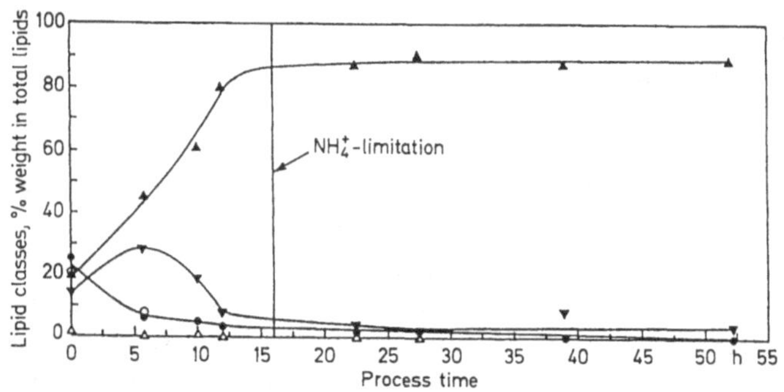

Fig. 3. Profile of the lipid classes of the extracted oil from *Rs. toruloides* at 35 °C and pH 7.5, in a controlled process (Biolafitte fermenter; 300 rpm; aeration rate = 0.6 vvm; 10 liters of medium; chemical antifoam addition). ▲ Triglycerides; △ free fatty acids; ▼ sterols; ● phospholipids; ○ glycolipids

4 Effects on the Fatty Acid Composition of Intracellular Microbial Lipids

Since most microorganisms have similar fatty-acid distribution, unsaturation of the fatty acids becomes a useful tool for comparison. Many factors affect the saturation of the fatty acids of the total oil produced, but results in the literature are highly contradictory.

On the one hand, the degree of unsaturation has been shown to increase if the following conditions are met: a) increased degree of unsaturation of a fatty-acyl supplement in the medium [11,84–86], or b) increased saturation of *n*-alkanes serving as C-source [87–89], or c) lower growth temperature [23,57,90], or d) increased aeration of the medium [84,91,92], or e) higher pH [7,92], or f) ascus formation [79], or g) pantothenic acid deficiency [80].

The above factors however do not constitute the general rule for all yeasts. It is generally recognized that a decrease in growth temperature results in the increase in the relative content of unsaturated fatty acids in the cellular lipids of almost all organisms, both procaryotes and eucaryotes. It is said that the cellular membrane must retain its fluidity at lower than optimum growth temperatures, hence the membrane phospholipids become more unsaturated. However, exceptions to this general rule exist. For example, *Saccharomyces cerevesiae* [83] showed little change in the degree of unsaturation of its lipids with a change in the temperature of batch or continuous incubation. But can one expect a similar rule for oleaginous microorganisms in which the main lipids, long after nitrogen limitation, are storage triglycerides (recall Fig. 3)? Most probably, since the individual fatty acids are synthesized first and are then incorporated to either phospholipids or triglycerides through the common intermediate phosphatidic acid [81,82]. It is indeed the case for *C. utilis* [93], a weak oleaginous yeast, as well as for *Rh. glutinis* [34] and *Rs. toruloides* [69], although not at every pH of incubation for the latter yeast. The main lipids of *C. lipolytica* grown in nitrogen-sufficient conditions were phospholipids and they followed the expected behaviour [94].

Exceptions also exist with oleaginous microorganisms. The absence of change in the unsaturation of lipids with temperature was found in *Cr. albidus* var. *albidus*[95], *Candida* NCYC 911[39] and *Rh. gracilis*[10]. Turcotte[69] found that a decrease of pH from 7.5 to 4.5 resulted in slightly more saturated lipids being formed at lower temperatures. Goulet[34] also found increased synthesis of unsaturated fatty acids at higher temperatures, as well as at lower ones, but showed that the fatty acid responsible for the unsaturation was C 18:2 at the higher end whereas C 18:1 and C 18:3 were present at the lower range.

The effect of temperature, however, might be masking the effect of a more important parameter, oxygen concentration. From thermodynamic considerations, a given aeration rate results in a higher dissolved oxygen (DO) content in a medium at lower temperature. Specifically, the ratio of DO at 10 °C to that at 35 °C is about 1.1. Coupled to that is the slower growth of the microorganisms at lower temperature, thus requiring less oxygen from the medium and resulting in an even larger ratio of DO. Since oxygen is involved in the biosynthesis of unsaturated fatty acids from saturated ones, it is expected that more unsaturation of the lipids would be found with increased aeration. Although many studies found this to be true[84,91,92] little effect on the degree of saturation of the lipids of *Rh. gracilis* and *Candida* NCYC 911[10,39] were found when those yeasts were grown in continuous culture under nitrogen-limiting conditions.

Kessell[7] argued that no significant changes occurred in the fatty acid composition of the lipid when nitrogen is exhausted from the medium. This result contradicts the findings of Dawson and Craig[76], of Goulet[34], of Hansson and Dostálek[95] but agrees with ours[69]. In the experiments of Dawson and Craig[76] with *C. utilis*, and those of Goulet[34] with *Rh. glutinis*, batch growth in a nitrogen-limited medium resulted in opposite patterns in the fatty acid distribution: *C. utilis* mainly increased its C 18:2 and C 18:3 while slightly decreasing its C 16:0 and C 16:1, its C 18:0 remained unchanged but C 18:1 reached a maximum sometime after nitrogen was exhausted, then fell close to its starting value. In *Rh. glutinis*, C 18:2 and C 18:3 are drastically reduced after nitrogen depletion, but C 16:0, C 18:0 and C 18:1 are slightly increased. Hansson and Dostálek[95] found that the fatty acids of the oil of *Cr. albidus* var. *albidus* varied most during growth, whether nitrogen was limiting or not (although the differences in composition varied more when excess nitrogen was present); while C 18:1 first decreased then increased to reach a plateau when glucose was depleted, C 18:2 following in exactly the inverse directions; C 16:0, C 18:0 and C 18:3 varied only slightly. Working with *Candida* NCYC 911, Gill et al.[29] reported that the degree of unsaturation of the fatty acids markedly decreased as the dilution rates of a carbon-limited medium increased, but that with nitrogen-limitation, the reverse trend was true. However, exactly the inverse situation was reported by Yoon and Rhee[38,77] with *Rh. glutinis*.

Rs. toruloides investigated in batch mode at combinations of temperature and pH[69] showed major changes in fatty acid distribution prior to nitrogen depletion only. The fatty acid, C 14:0, remained approximately constant around the 1% level; C 16:0 increased slightly (5 to 15%) up to its steady-state value whereas C 16:1 was low (typically 1%), if present at all; C 18:0 also usually increased slightly (about 5 to 10%) before levelling off to its equilibrium value. The most dramatic changes occurred in the profiles of the unsaturated fatty acids of the C 18 series. The C 18:1

acid decreased to a minimum while C 18:2 and C 18:3 increased simultaneously, then the reversed trend was observed. The concentration of the individual fatty acids present in the extracted lipids reached a steady-state value soon after nitrogen-depletion, in most cases. This reversal pattern in polyunsaturated fatty acid profiles was recurrent in all experiments performed; the conditions of pH and temperature seemed to mainly affect the width and the depth of the peaks, although in a minimal way. Thus, it seems likely that nitrogen and glucose catabolisms are responsible for the oscillating behaviour of oleic, linoleic and/or linolenic acids. The age of the culture was also shown to affect the temporal pattern of fatty acid distribution in other studies [23,89].

Interestingly enough, when linoleic acid (and/or linoleic acid) and oleic acid are the principal fatty acids to change in response to a certain operating condition, the patterns are usually in opposite directions [7,29,34,58,69,94,95]. The high concentration of linolenic acid in some microorganisms has attracted the interest of companies in Japan [96] and England [97] where commercial production of γ-linolenic acid on an industrial scale seems very promising.

Welch and Burlingame [98] reported a maximum of 2% fatty acids with more than 18 carbons in a wild-type of *S. cerevisiae* with C 26:0 being the most abundant. Long-chain fatty acids are of particular interest since elucidation of their biosynthesis could lead to large-scale industrial production. One such microbial process is reportedly taking place in Japan for the production of brassylic acid, a long-chain dicarboxylic acid, from the yeast *C. tropicalis* [99]. The production of arachidonic acid and eicosapentaenoic acid from strains of *Mortierella* also seems promising.

5 Extracellular Microbial Lipids

When certain yeasts are grown on agar plates, oil globules may appear on the surface of the solid medium. Although many investigators reported such findings, only a few tried to quantify the fatty acid composition of these extracellular lipids [9,100,101].

When cultivation occurs under optimum growth conditions, the more important fatty products (fatty acids, alcohols) never accumulate in the medium in any significant concentrations. Zalashko and Pidoplichko [102] found a negative correlation between the synthesis of intracellular lipids and the quantity of extracellular oil for eleven strains of *Rhodotorula*. Zalashko et al. [9] found that the extracellular lipids of the yeast *Rh. glutinis* contained no phospholipids or glycerides, as opposed to the intracellular fat of the same organism.

The effect of aeration on biosurfactant production was studied by Gradova et al. [91] for microorganisms grown on petroleum fractions. They found that: a) with moderate aeration, the maximum production of partially oxidized hydrocarbons occurred during the lag-phase of the culture; b) less aeration had a tendency to retard the growth of the biomass and certain metabolic reactions thereby resulting in the accumulation of intermediate products of oxidation (fatty acids), c) with more aeration, the hydrocarbons were more completely oxidized and the fatty acid content in the medium fell.

Secondary products of metabolism are also excreted by yeasts [25]. They are classified in four different types of compounds:

a) polyol fatty acid esters, in which saturated, unsaturated, and hydroxy fatty acids are joined by ester linkages to C_5 and C_6 polyols;
b) sophorosides of hydroxy fatty acids, in which saturated and unsaturated hydroxy fatty acids are linked glycosidically to the disaccharide sophorose;
c) acetylated sphingosines, in which all of the hydroxy and amino groups of C_{18}-phytosphingosine and C_{18}-dihydrosphingosine are acetylated;
d) C_{22} acids, in which trihydroxy and dihydroxyoxo fatty acids are acetylated.

Torulopsis bombicola produces a variety of oligo-glycolipid biosurfactants when incubated in nitrogen-limiting conditions [102] and a combined carbon source of vegetable oil (or fatty acids such as oleic, or methyl esters of stearic or oleic acids) and glucose [103–105]. Yeasts of the genus *Torulopsis*, *Candida* and other oleaginous microorganisms reportedly produce surface-active lipids on both alkanes and water-soluble compounds [19]. The chain length of the lipophilic moiety of the biosurfactant usually reflects that of the utilized alkane.

Although a strong correlation exists between the emulsification/solubilization of alkanes by extracellular lipids and of uptake of alkanes, the extremely high extracellular concentration of lipids is more likely the result of particular biotechnological process parameters, or of particular conditions of limitation, than a necessary metabolic reaction to alkane utilization [106]. In any cases, the extracellular lipid accumulation and the reason for excretion have not been elucidated up to now. Furthermore, the industrial production of biosurfactants on water-insoluble compounds is hampered by difficulties in separation of cells from media and subsequent isolation and identification of the surface-active material.

6 Concluding Remarks

It seems possible to tailor-make large quantities of triglycerides of microbial origin using the following guidelines:
a) choose an oleaginous (fat-producing) yeast (strains of *Candida*, *Lipomyces*, or *Rhodotorula*),
b) place in a nitrogen-limiting medium,
c) grow the cells in a single-stage batch manner for 3d or more, or at very low dilution rates in a chemostat mode, and possibly
d) control temperature at or slightly lower than that for maximum growth conditions and in a slightly acidic medium.

If a particular fatty acid profile is required, the selection of the yeast could be based on reported fatty acid distributions from the literature as long as these are steady-state values (obtained some time after nitrogen-depletion). It might also be possible to alter the ratio of C 18:1 to C 18:2 (and/or to C 18:3) by feeding steady-state cultures with nitrogen (and carbon, if it reached too low a level) and harvesting the cells prior to the depletion of this second input of nitrogen.

It would be of a great advantage to study the correlations between variables; e.g., would temperature affect lipid production in the same manner at various growth rates *and* aeration rates, *or* are there synergistic effects for some specific combination of these parameters?

In most studies described in this review, batch cultures are used with an amount of nitrogen much lower than that required for maximum growth. When the nitrogen level is very low, the cells are forced to utilize the excess sugar present and to transform it into storage lipid. Only a few authors [39] investigated lipid production using the cells' own growing pattern: a first stage to produce maximum biomass, and a second one to store the lipid. They came to the conclusion that a two-stage continuous system offered no practical advantages over a single-stage system.

Since lipid production in continuous systems seems to occur with long residence times (typically larger than 50 h), it is preferable to work in a batch mode. As the average residence time in the bioreactor implies, the microbes behave essentially as if they were reacting to a batch environment.

Ratledge and his co-workers [107–118] studied the effect of the enzymes involved in lipid biosynthesis. The elucidation of the biochemical pathways could open a new era of research in genetically engineered fattened cells.

7 Nomenclature

[ ] concentration, M or g l^{-1}
rpm revolutions per minute, min^{-1}
vvm volume of air per volume of solution per minute
μ specific growth rate, h^{-1}

8 References

1. Ratledge C (1976) In: Food from waste Birch GG, Parker KJ, Worgan JT (eds) Applied Science, London, p 98
2. Fedeli E, Jacini G (1971) Adv. Lipid Res. 9: 335
3. Jurriens SG (1968) In: Boekennogen HA (ed) Analysis and characterization of oils, fats and fat products, Interscience, New York, vol 12, p 217
4. Ratledge C (1978) Lipids and fatty acids. In: Rose AH (ed) Economic microbiology, Academic, New York, vol 2, p 263
5. Nylen P (1965) Modern surface coatings, Interscience, New York
6. Mumma RO, Sekura RD, Fergus CL (1971) Lipids 6(8): 584
7. Kessell RHJ (1968) J. Appl. Bacteriol. 31: 220
8. Hartman L, Hawke JC, Shortland FB, di Menna ME (1959) Arch. Biochem. Biophys. 81: 346
9. Zalashko MV, Romanova LV, Gerbeda VV, Pidoplichko GA (1979) Microbiol. 48(2): 181
10. Choi SY, Ryu DDY, Rhee JS (1982) Biotechnol. Bioeng. 24(5): 1165
11. Watson K, Rose AH (1980) J. Gen. Microbiol. 117(1): 225
12. Thorpe RF, Ratledge C (1972) J. Gen. Microbiol. 72: 151
13. Tatsumi C, Hashimoto Y, Terashima M, Matsuo T (1977) U.S. Patent, No. 4,032,405
14. Kammann KP Jr, Phillips AI (1985) JAOCS 62(5): 917
15. Carlson KD, Chang SP (1985) JAOCS 62(5): 934
16. Fulmer RW (1985) JAOCS 62(5): 926
17. Sonntag NOV (1985) JAOCS 62(5): 928
18. Cooper DG, Zajic JE (1980) Adv. Applied Microbiol 26: 229
19. Haferburg D, Hommel R, Claus R, Kleber HP (1986) Adv. Biochem. Eng./Biotechnol. 33: 53
20. Fujii T, Tonomura K (1971) Agri. Biol. Chem. 35(8): 1188
21. Dewitt S, Ervin JL, Howes-Orchison D, Dalietos D, Neidleman SL, Geigert J (1982) JAOCS 59(2): 69

22. Enebo L, Anderson LG, Lundin H (1946) Arch. Biochem. 11: 383
23. Moon NJ, Hammond EG (1978) JAOCS 55(10): 683
24. Ratledge C (1982) Microbial oils and fats: an assessment of their commercial potential. In: Bull MJ (ed) Prog. Ind. Microbiol. Elsevier, Amsterdam, vol 16, p 119
25. Spencer JFT, Spencer DM, Tulloch AP (1978) In: Rose AM (ed) Economic microbiology, Academic, New York, p 523
26. Uzuka Y, Kanamori T, Koga T, Tanaka K, Naganuma T (1975) J. Gen. Appl. Microbiol. 21(3): 157
27. Gierhart DL (1984) U.S. Patent, No. 4,485,172
28. Atamanyuk DI, Vakar LT (1975) Lipidy Gribov: 98
29. Gill CO, Hall MJ, Ratledge C (1977) Appl. Environ. Microbiol. 33(2): 231
30. Ratledge C, Hall MJ (1979) Biotechnol. Letters. 1: 115
31. Roy MK, Vadalkar K, Baruah B, Misra U, Bhagat SD, Baruah JN (1978) Indian J. Exp. Biol. 16(4): 511
32. Allen LA, Barnard NH, Fleming M, Hollis B (1964) J. Appl. Bacteriol. 27(1): 27
33. Gorodnyanskaya LI, Zvyagintseva IS (1978) Microbiol 47(3): 369
34. Goulet J (1975) Ph.D. Thesis, McGill University, Montreal, Canada
35. Ogata K, Kaneyuki H, Kato N, Tani Y, Yamada H (1973) J. Ferment. Technol. 51(4): 227
36. Park SO (1974) Hanguk Nonghwa Hakhoe Chi 17(2): 93
37. Pitryuk IA, Zvyagintseva IS, Bab'eva IP (1975) Microbiol. 44(5): 734
38. Yoon SH, Rhee JS (1983) JAOCS 60(7): 1281
39. Hall MJ, Ratledge C (1977) Appl. Environ. Microbiol. 33(3): 577
40. Hansson L, Dostálek M (1986) JAOCS 63(9): 1179
41. Maas-Forster M (1955) Archiv für Mikrobiologie 22: 115
42. Nielsen N, Nilsson NG (1953) Acta Chem. Scandinavica 7: 984
43. Evans CT, Ratledge C (1984) J. Gen. Microbiol. 130: 1693
44. Evans CT, Ratledge C (1984) J. Gen. Microbiol. 130: 1705
45. Glatz BA, Floetenmeyer MD, Hammond EG (1985) J. Food Prot. 48(7): 574
46. Atamanyuk DI, Borisova TA, Tsygulya TE (1978) Izv. Akad. Nauk Mold SSR, Ser. Biol. Khim. Nauk 5: 87
47. Picataggio SK, Smittle RB (1979) Eur. Pat. Appl. 277
48. Atamanyuk DI, Vakar LI (1976) Izv. Akad. Nauk Mold. SSR, Ser. Biol. Khim. Nauk 4: 48
49. Simard RE, Busque G, Riel RR (1973) Can. Inst. Food Sci. Technol. J., 6(1): 32
50. Nojiri M, Kakutani K, Uedono S, Uenakai K, Matsumoto M (1979) Ger. Offen., 2, 728: 353
51. Dostálek M (1986) Appl. Microbiol. Biotechnol. 24(1): 19
52. Maugenet J (1973) C.R. Seances Acad. Agr. Fr. 59 (7): 481
53. Floetenmeyer MD, Glatz BA, Hammond EG (1985) J. Dairy Sci. 68(3): 633
54. Davies RJ (1983) Rep.-N.Z., Dep. Sci. Ind. Res., Ind. Process Div., IPD/TSO: 2011
55. Steinberg M, Ordal Z (1954) J. Agric. Fd. Chem. 2: 873
56. Kates M, Baxter RM (1962) Can. J. Biochem. Physiol. 40: 1213
57. Thorpe RF, Ratledge C (1973) J. Gen. Microbiol. 78: 203
58. Enebo L, Iwamoto H (1966) Acta Chem. Scand. 20(2): 439
59. Guarneri JJ, Combs TJ, Pisano MA (1987) Dev. Ind. Microbiol. 20: 641
60. Heide S (1939) Arch. Mikrobiol. 10: 135
61. Raaf H (1941) Arch. Mikrobiol. 12: 131
62. Smedley-MacLean I (1922) Biochem. J. 16: 370
63. Smedley-MacLean I, Hoffert D (1923) Biochem. J. 17: 720
64. Pan SC, Andreasen AA, Kolachov P (1949) Arch. Biochem. 23: 419
65. Spotholz CH, Lichtfield JH, Ordal ZJ (1956) Appl. Microbiol. 4: 285
66. Combs TJ, Guarneri JJ, Pisano MA (1980) Dev. Ind. Microbiol. 21: 373
67. Cocucci MC, Belloni G, Gianani L (1975) Arch. Microbiol. 105: 17
68. Ratledge C, Hall MJ (1977) Appl. Envir. Microbiol. 34(2): 230
69. Turcotte G (1987) Ph. D. Thesis, The University of Western Ontario, London, Ontario
70. Harrison DEF (1973) CRC Critical Reviews in Microbiology 2: 185
71. Ishizaki A, Shibai H, Hirose Y (1974) Agr. Biol. Chem. 38(12): 2399
72. Wimpenny JWT (1969) Biotechnol. Bioeng. 11: 623
73. Shibai H, Ishizaki A, Kobayashi K, Hirose Y (1974) Agr. Biol. Chem. 38(12): 2407

74. Akashi K, Ikeda S, Shibai H, Kobayashi K, Hirose Y (1978) Biotechnol. Bioeng. 20(1): 27
75. Wimpenny JWT, Necklen DK (1971) Biochim. Biophys. Acta. 253
76. Dawson PSS, Craig BM (1966) Can. J. Microbiol. 12: 775
77. Yoon SH, Rhee JA (1983) Process Biochem., October 2
78. Ghose TK, Chotani GK, Ghosh P, Sahai V (1987) Ann. N.Y. Acad. Sc. 506: 459
79. Illingworth RF, Rose AH, Beckett A (1973) J. Bacteriol. 113(1): 373
80. Furukawa Y, Kimura S (1972) J. Vitaminol. 18(4): 213
81. Lehninger AL (1972) Biochemistry, Worth Publishers, Inc., Ch. 23 New York
82. Vance DE, Vance JE (eds) (1985) Biochemistry of lipids and membranes, Benjamin/Cummings Menlo Park, California
83. Hunter K, Rose AH (1972) Biochim. Biophys. Acta, 260(4): 639
84. Jollow D, Kellerman GM, Linnane AW (1968) J. Cell Biol. 37: 221
85. Mudd JB, Saltzgaber-Muller J (1978) Arch. Biochem. Biophys. 186(2): 359
86. Glatz BA, Hammond EG, Hsu KH, Baehman L, Bati N, Bednarski W, Brown D, Floetenmeyer MD (1984) AOCS Monogr., 11 Biotechnol. Oils Fats Inc.: 163
87. Harries PC, Ratledge C (1969) Chem. Ind., 45: 582
88. Hornei S, Kohler M, Weide H (1972) Zeitschrift für Allgemeine Mikrobiologie, 12(1): 19
89. Volfova O, Pecka K (1973) Folia Microbiol.: 18
90. Rattray JB, Schiberi M, Kidby DK (1975) Bacteriol. Rev. 39(3): 197
91. Gradova NB, Kruchkova AP, Rodionova GS, Mikhaylova VV, Dikanskaya EM (1969) Biotech. Bioeng. Symp. 1: 99
92. Zalasko MV, Gurinovich ES, Obraztsova NV, Koroleva IF, Bogdanovskaya ZhN, Andreevskaya VD, Andriyashina SS (1974) Khimiya Tverdogo Topliva, 3(5): 146
93. Brown CM, Rose AH (1969) J. Bacteriol. 99(2): 371
94. Kates M, Paradis M (1973) Can. J. Biochem. 51: 184
95. Hansson L, Dostálek M (1986) Appl. Microbiol. Biotechnol. 24(3): 187
96. Suzuki O (1987) Proceed. World Conf. Biotechnol. Fats Oils Ind., Hamburg, Germany
97. Chementator: (1988) Chem. Eng., March 28: 13
98. Welch JW, Burlingame AL (1973) J. Bacteriol. 115(1): 464
99. Uemura N (1987) Proceed. World Conf. Biotechnol. Fats Oils Ind., Hamburg, Germany
100. Kaneko H, Hosohara M, Tanaka M, Itoh T (1976) Lipids, 11(12): 837
101. Stodola FH, Deinema MH, Spencer JFT (1967) Bacteriol. Rev. 31(3): 194
102. Zalashko MV, Pidoplichko GA (1972) Microbiol. 41(4): 581
103. Asmer HJ, Lang S, Wagner F, Wray V (1987) Proceed. World Conf. Biotechnol. Fats Oils Ind. Hamburg, Germany
104. Cooper DG, Paddock DA (1984) Appl. Envir. Microbiol. 47(1): 173
105. Inoue S (1987) Proceed. World Conf. Biotechnol. Fats Oils Ind., Hamburg, Germany
106. Duvnjak Z, Cooper DG, Kosaric N (1982) Biotechnol. Bioeng. 24: 165
107. Botham PA, Ratledge C (1979) J. Gen. Microbiol. 114: 361
108. Boulton CA, Ratledge C (1980) J. Gen. Microbiol. 121: 441
109. Boulton CA, Ratledge C (1981) J. Gen. Microbiol. 127: 169
110. Boulton CA, Ratledge C (1981) J. Gen. Microbiol. 127: 423
111. Boulton CA, Ratledge C (1983) J. Gen. Microbiol. 129(9): 2863
112. Evans CT, Ratledge C (1983) Lipids 18(9): 630
113. Evans CT, Ratledge C (1985) Can. J. Microbiol. 31(11): 1000
114. Evans CT, Ratledge C (1985) Biotechnol. Genet. Eng. Rev. 3: 349
115. Evans CT, Ratledge C (1985) Can. J. Microbiol. 31(5): 479
116. Evans CT, Ratledge C (1985) Can. J. Microbiol. 31(9): 845
117. Whitworth DA, Ratledge C (1975) J. Gen. Microbiol. 90: 183
118. Whitworth DA, Ratledge C (1975) J. Gen. Microbiol. 88: 275

The Use of Calorimetry in Biotechnology

Urs von Stockar and Ian W. Marison
Institute of Chemical Engineering, Swiss Federal Institute of Technology (EPFL)
CH-1015 Lausanne, Switzerland

1 Introduction ... 94
2 Calorimetric Measuring Principles ... 97
2.1 Microcalorimeters ... 97
2.1.1 Adiabatic Calorimeters ... 97
2.1.2 Compensation Calorimeters ... 98
2.2 Flow Microcalorimeters ... 100
2.3 Isothermal Reaction Calorimeters ... 102
2.3.1 Dynamic Calorimetry ... 102
2.3.2 Continuous Calorimetry ... 103
2.3.3 Heat Flux Calorimetry ... 104
3 Theory ... 106
3.1 Simultaneous Elemental and Enthalpic Balance for Cellular Growth Processes ... 106
3.2 Elemental and Enthalpic Contents of Substrate and Products and the Role of the Degree of Reduction ... 108
3.3 Predicting Important Heat Yields ... 111
3.4 Modelling Special Cases ... 113
4 Applications ... 115
4.1 Measurement of Microbial Growth ... 115
4.2 Biomass Yield ... 119
4.3 Maintenance Energy ... 122
4.4 Nature of Substrate and Growth Efficiency ... 124
4.5 Oxygen Up-take ... 127
4.6 Product Formation ... 128
5 Conclusions ... 131
6 Symbols ... 132
7 References ... 134

After a short review of the development of biological calorimetry and a discussion of the instrumentation and the measuring principles that have been applied to study heat generation by microbial cultures and other cellular systems, this article demonstrates how heat evolution depends on growth, biomass yield, maintenance metabolism, nature of substrate, energetic efficiency of growth, oxygen uptake, and product formation. Theoretical considerations are used together with experimental evidence to explain the nature of these relationships and the underlying reasons. The possibilities of exploiting these relationships in order to gain information on any of the above factors by applying calorimetric measurements to biotechnology have been emphasized and developed throughout this paper.

1 Introduction

The potential role of calorimetry in biotechnology is founded on the fact that all living organisms of all kinds and without exception invariably produce heat. The phenomenon of life itself is intimately and inherently tied to the concept of heat dissipation. Athermal or even endothermic growth processes are unthinkable as they would violate the second law of thermodynamics.

It is therefore perhaps no coincidence that the first really operative calorimeter ever described in the literature was more devoted to studying biological heat generation than heat production by chemical reactions. In 1780, Lavoisier and de Laplace placed a guinea pig into an adiabatically insulated bucket which was completely lined inside with ice (Fig. 1). They observed the metabolic heat generation by collecting the melted ice and measuring its amount [1]. Although the body weight of the guinea pig used by Lavoisier and de Laplace was not noted in the report, we may conclude from their measurements that guinea pigs may release something on the order of 3 W per kg of body weight (Table 1). As can be seen from

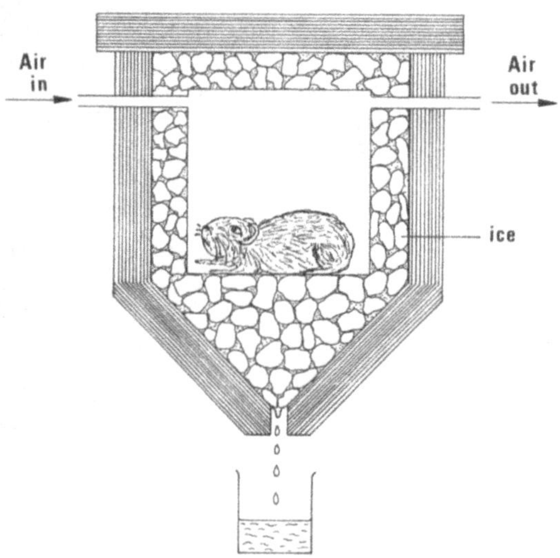

Fig. 1. Early ice calorimeter experiments of Lavoisier et de Laplace

Table 1. Biological heat generation rates

Organism	Author	Heat generation rate		
		Watts per		
		Organism	kg living mass	kg dry mass
Man	Jéquier, [2]	100.0	1.3	—
Guinea pig	Lavoisier & de Laplace, [1]	3.5	3	—
Resting yeast culture	Birou & von Stockar, [3]	—	5	25
Yeast culture, growing exponentially	Marison & von Stockar, [4]	—	250	1330

Table 1, guinea pigs are obviously a more "high-powered" species than mankind, whose energetic output amounts to a mere 1.3 W kg^{-1} according to more recent direct measurements made at the University of Lausanne [2]. Microorganisms are even more energetic: resting yeast cultures dissipate about 25 W kg^{-1} dry biomass [3] but easily step their power dissipation up to well over 1 kW kg^{-1} when they start to grow exponentially [4].

These figures hint to the fact that considerable power may be dissipated in the form of heat as a result of life processes, yet they are small in comparison with the heat production that may be attained in chemical reactions. This is probably the main reason why biological calorimetry, after having made a promising start owing to Lavoisier, somewhat trailed behind development of chemical calorimetry. Indeed, the faint heat production rates accompanying biological phenomena are so hard to detect that biologists and microbiologists often do not even take note of them.

Biological calorimetry lay dormant for the better part of a century. We owe the first microbiological measurement to Dubrunfaut, who published in 1856 a "Note sur la chaleur et le travail produits par la fermentation vineuse" [5]. He solved the problem of measuring small heat outputs by working with a huge vat containing 21400 liters of culture fluid, in which 3.5 t of sugar were degraded.

The first direct laboratory measurement of microbial heat generation was not performed until another half-century later and is due to Rubner [6–8], although some researchers had tried somewhat earlier to determine the enthalpy difference of culture media before and after growth using a Berthelot-type bomb calorimeter [9].

The need for highly sensitive instrumentation in order to quantify the feeble biological signal spurred the development and perfection of modern microcalorimeters [10–11]. Due to their extraordinary sensitivity, the overwhelming majority of studies on microbial heat production were effected in microcalorimeters up through the late seventies. The lack of agitation and aeration in the early versions of such microcalorimeters limited their applicability to studies of anaerobic cultures, but later designs affording at last some level of mixing and oxygenation permitted studies of modestly aerobic cultures [12–14]. Thus, classical biological microcalorimetry became, at last, a valued tool for microbiology and was used by a number of researchers for finger-printing microbial growth patterns, for the study of antibacterial drug metabolism, for fundamental studies of microbial growth energetics, and the like [15–18].

Nevertheless, all of this microcalorimetric research made hardly any impact on biotechnology, bioprocess development or bioprocess engineering. In the early seventies, it was, for instance, still quite impossible to design the cooling facilities for industrial scale processes based on a calorimetric experiment, and engineers usually had to resort to crude estimations based on energy balances. The fact that most microcalorimeters can neither be efficiently aerated nor agitated simply precludes their use for the highly aerobic processes of industrial interest. Because no process variable other than temperature can usually be controlled, the growth conditions in a microcalorimeter are very different, (usually much worse) from those prevailing in technical reactors. Furthermore, the fact that usually no process variable other than heat flux can be measured directly in the medium hampers the quantitative interpretation of the observed heat fluxes and their correlation with other rate information. In an excellent review of the technical difficulties and pitfalls encountered with microcalorimeters, Wadsö [19] wrote in 1980: "For systems with a very high oxygen consumption, such as many aerobic microbial growth processes of biotechnical interest, there appears, as yet, to be no suitable calorimeter available."

Progress in this direction has however been made during recent years. Two research groups proposed new microcalorimeter designs, which paralleled the design of laboratory bioreactors quite closely. They equipped their micro-vessels with miniaturized stirrers and in some instances with air spargers [20-23]. As early as 1968, Cooney et al. [24] pioneered the completely novel development of calorimeters at the bench scale. Three different types of bench scale calorimeters for use in biotechnology have been described in the literature to date [24-31]. They all consist of a laboratory bioreactor with a nominal volume comprising between 2 and 14 l, which can be operated much like a standard laboratory reactor, while featuring the necessary equipment for measuring the heat generated by the microbial culture in situ. It is obvious that this new form of calorimetry, which has become known in chemical engineering literature as 'Isothermal Reaction Calorimetry' [32] is especially well suited for quantitative investigations of microbially generated heat-fluxes under technically realistic processing conditions. Their main disadvantage, a relatively low sensitivity, can, in part, be overcome by using a novel "heat-flux" measuring principle [30].

Judging from published results, the more recent experimental techniques developed in the last decade will probably expand the usefulness of calorimetry in biotechnology from the traditional field of utilization of microcalorimetry to incorporate more quantitative process-engineering related applications, some of which are summarized in Table 2.

At the laboratory scale, calorimetry could not only be used in order to determine process cooling requirements, but it might reveal itself as a valuable tool for bioprocess development work. Due to the continuous and unspecific nature of the heat signal, the microbial activity can constantly be monitored, special events or

Table 2. Main uses of calorimetry in biotechnology

1. Process development work
   - Experimental determination of process cooling requirements
   - Monitoring biological activity as a function of process conditions, detecting unsuspected special events and anomalies such as limitations, product formation, diauxic growth, etc.
   - Medium optimization: identification of limiting factors and inhibitors
   - Rapid and quantitative characterization of growth kinetics
   - Investigation of limiting mass transfer phenomena
2. Bioprocess and bioreactor control
   - Use of heat dissipation measurement at production scale as an on-line "probe" for indirect determination of
     - Biomass concentration and growth
     - Product formation
     - State of culture
   - Use of on-line heat dissipation measurement at production scale together with other on-line signals for computer-aided control of bioreactor
3. Research
   - Enzyme-level kinetics of culture
   - Induction-repression kinetics in continuous culture
   - Growth energetics and biothermodynamics

anomalies can immediately be pinpointed, and microbial growth kinetics can be quantitatively determined. Quantitative heat dissipation measurements at the laboratory scale will undoubtedly also be exploited for basic research.

At the industrial scale, heat generation measurements may hold considerable potential for process control purposes. Computer control of bioprocesses, whose importance for biotechnology is steadily increasing, is often limited by the number of relevant process variables that can be measured on-line and used as input signals for the computer. Suitable probes for such data acquisition are available only for a limited number of process variables. The heat dissipation rates can, however, be measured on-line in a large bioreactor without resorting to sophisticated on-line probes. The accuracy of such measurements improves with the scale due to the decreasing surface-to-volume ratio. On-line measurements of microbially generated heat fluxes could therefore be used along with other on-line data in order to aid bioreactor control, provided that the heat signal can be interpreted and its relation to other relevant process variable is known.

2 Calorimetric Measuring Principles

It is not within the scope of the present work to discuss all of the types of calorimeter which are available for investigations of chemical and biochemical reaction systems. Instead the structure and measurement principles of the major calorimeter systems used in biotechnology will be discussed, with particular emphasis on the advantages and disadvantages of each system for specific applications. For more detailed information on calorimetry systems, the reader is referred to one of the many reviews which are available on this subject [19, 33–39].

Calorimeters may be defined on a number of criteria relating to specific measurement principles. With regard to application in biotechnology, calorimetric devices may be arbitrarily divided into micro- and macro- (or bench scale) calorimeters. This classification is based purely on the size of the calorimetric measurement cell, although it reflects a difference in the principle of operation and relative sensitivity of the systems. These groups may be further divided into adiabatic, compensation and modified compensation calorimeters.

2.1 Microcalorimeters

2.1.1 Adiabatic Calorimeters

The principle of adiabatic calorimeters is to minimize the heat loss to the surroundings such that any heat liberated by the process under study results in an increase in the temperature of the reaction mixture.

Under ideal conditions there is no net heat exchange between the measuring system and the surroundings. If these conditions are not completely met, it will result in a temperature gradient between the reaction mass and the surroundings. In this case, the system is defined as an isoperibol calorimeter and it is necessary to calibrate the system to define this heat loss [40].

Such calorimetric systems have found relatively little application in biotechnology since the reaction is accompanied by a temperature change which can seriously affect the metabolism of the cells.

2.1.2 Compensation Calorimeters

Compensation calorimeters have been described by a variety of alternative names including heat-conduction calorimeters, heat-flux or heat-flow calorimeters, thermopile conduction calorimeters, etc. They are by far the most commonly used systems operating on similar principles but achieving the temperature regulation of the system by a variety of methods. Such systems are commercially available from Setaram (France), Beckman Instruments (USA) and LKB Produkter (Sweden).

In an ideal compensation or heat conduction calorimeter, there is a controlled transfer or flow of heat from the calorimeter vessel to the surroundings (heat sink). This control is usually achieved by surrounding the calorimeter vessel with a thermopile wall (Fig. 2). The temperature difference over the thermopile wall gives rise to a voltage signal which is described by:

$$V = k \frac{dQ}{dt} \tag{1}$$

The time constants of such calorimeters range from approximately 100 seconds to several minutes. The fast-response calorimeters are generally those with the smallest reaction volumes, which accounts for the predominance of microcalorimeter systems.

In practice, a significant part of the total heat flow is not through the thermopiles but through the insulation material, cables and air gaps between the thermopiles and the heat sink. These effects, together with the time constant of the apparatus and sensitivity, can be determined by electrical or chemical calibration but can result in signals which do not closely resemble the true reaction kinetics, particularly with fast reactions.

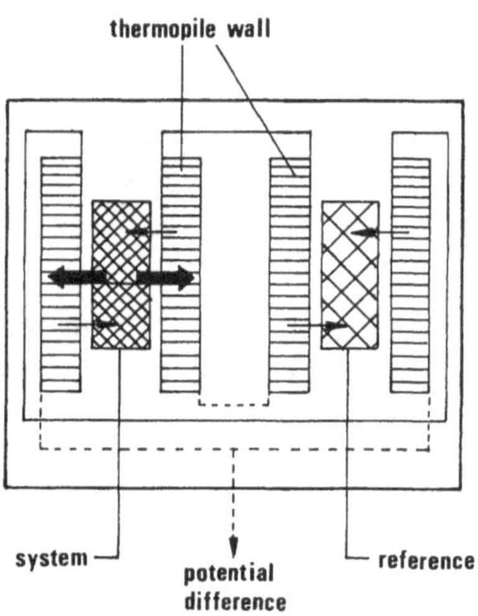

Fig. 2. Schematic representation of a section through a 'Calvet-type' compensation calorimeter

The most widely used heat conduction calorimeters are those developed by Calvet based on the initial design of Tian [10,41–43]. A typical Calvet-type calorimeter as marketed by Setaram (France) has a sensitivity of the order of 1 μW and generates a relatively stable heat signal. In this case, between 2 and 4 calorimetric vessels are held in an aluminium block (heat sink). The reaction volume is between 15 to 100 ml and some of the calorimetric vessels may be gently stirred and materials added (injected) to the vessels [42,43]. Versions of this basic structure may be agitated by rotating the whole block.

Frequently the voltage generated is a differential signal between a calorimetric cell in which the reaction proceeds compared with a reference vessel. Such twin-heat conduction calorimeters have a small volume (1 to 5 ml) and high sensitivity 1–2 μW. Such devices have been used with a wide range of biological systems including microbial and animal cells [44–46].

The major limitations involved in the use of compensation calorimeters for the study of microbial growth processes are caused by the extreme sensitivity of these "micro-calorimeters", the signals of which are subject to significant interference by certain process operations. Thus, the culture held in the measurement cell cannot be efficiently agitated, which may result in sedimentation of the microbial cells and the formation of local temperature and nutrient/product concentration gradients. Efficient aeration of the culture is practically impossible which limits application to anaerobic processes. Control of the culture pH is difficult to achieve due to the large heat signals generated through heats of solution and neutralization caused by acid or base addition. It is difficult to add medium continuously to, or remove samples or medium from the measurement cell without seriously affecting the heat signal. This effectively precludes the use of such methods for continuous culture studies or where samples need to be removed for analysis of medium components, biomass concentration, product concentration and so on.

In order to alleviate such problems, Ishikawa et al. [20] have developed a twin-type heat conduction (compensation) calorimeter (Figs. 3 and 4). In this system two stain-

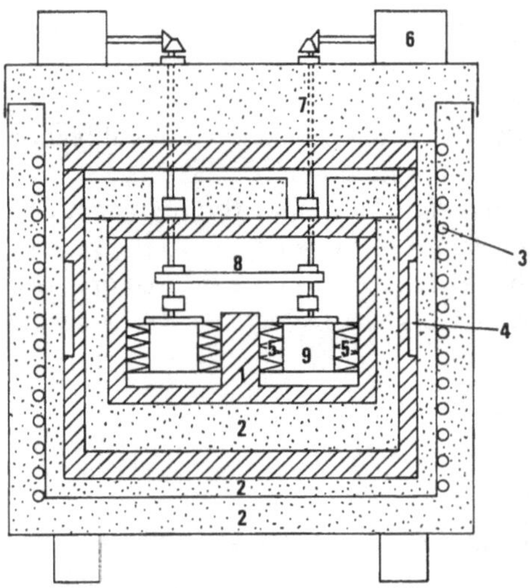

Fig. 3. Schematic representation of the twin, heat conduction calorimeter of Ishikawa et al. [20]. (*1*) heat sink (*2*) styrofoam insulating layer (*3*) cooling layer (*4*) thermomodule (*5*) thermopile (*6*) motor (*7*) shaft (*8*) aluminium bar (*9*) calorimeter vessel (Re-drawn with permission from [20])

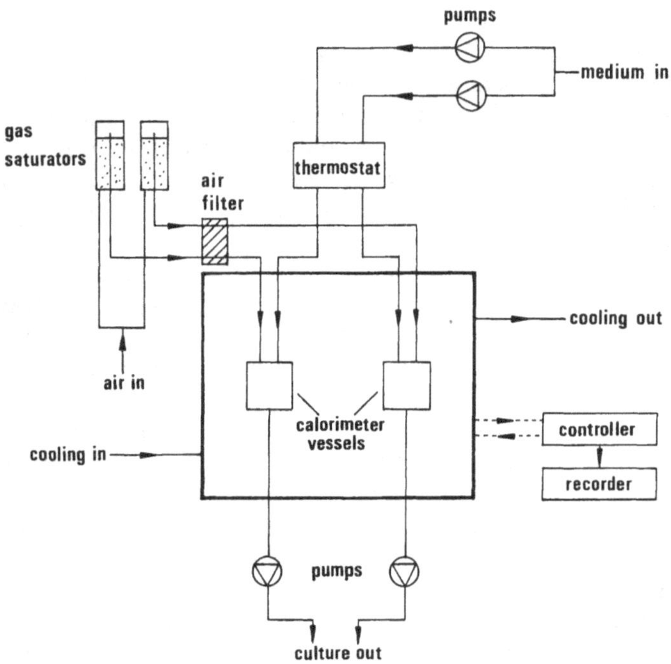

Fig. 4. Schematic representation of twin, heat conduction calorimeter system as used by Ishikawa et al. [20]

less steel 30 ml working-volume reaction vessels, equipped with agitation and aeration systems, are each surrounded by 4 thermopiles consisting of a total of 140 copper-constantin thermocouples. Each of the vessels is filled with culture medium and one is inoculated with microorganisms (culture vessel). Through careful insulation, presaturation, and humidification of the airstream, and by oxygen-enriching the air-stream to reduce the air flow rate into the vessel, the sensitivity of this calorimeter system, defined as the ratio of voltage output of the calorimeter to a known energy supplied using an electrical calibration heater, was reported to be 2 μV $(mW)^{-1}$ [20]. This system has been successfully used to study both batch and continuous cultivation of the bacterium *E. coli* under a range of nutrient limitations [20,21,47].

A major limitation of this and similar methods [22,48] is that the pH cannot be controlled and samples cannot be effectively removed for analysis. Consequently, a bioreactor with similar characteristics to the calorimeter vessels has to be operated in parallel for the determination of growth parameters such as biomass, substrate and product concentrations.

2.2 Flow Microcalorimeters

A further group of modified compensation calorimeters are those typified by twin conduction calorimeters through which the culture sample flows. The microbial culture is grown in a bioreactor in either batch or continuous mode and samples of the culture are continuously pumped to the calorimeter measuring cell (Fig. 5).

Monk and Wadsö [49] were among the first to develop such a flow calorimeter for precise biological measurements. This design has been improved upon and forms the basis of the 'Biological Activity Monitor' commercialized by LKB Produkter, Sweden. The instrument is composed of two calorimeter flow-through vessels (Fig. 6 [50]) surrounded by semiconducting thermocouple plates which, in turn, are surrounded by small aluminum blocks which are in thermal contact with the main calorimeter block. The whole heat sink arrangement is insulated and immersed in a thermostated water bath.

Samples are pumped through small diameter (1 mm internal diameter) gold tubes held between copper plates in the central heat exchange unit before entering the calorimetric measuring unit. The volume of the tubes and measurement vessel is approximately 0.5 ml. The heat-exchanger is to ensure that the sample is brought to a standard temperature before measurement of the heat generation rate which is a differential voltage signal between the reaction and reference vessels.

Various types of calorimetric vessel involving flow through cells, ampoules and perfusion devices have been developed for studies of aerobic bacterial growth and blood cells [50].

The main problems associated with such systems are due to the transfer time of the sample from the bioreactor to the measurement cells, which takes a minimum of about 2 min, and problems of cell adhesion to the walls of the gold tubes in the measurement cell [15]. The transfer time could result in modifications in the metabolism of the microbial cells caused by oxygen or substrate limitation [15]. A further problem

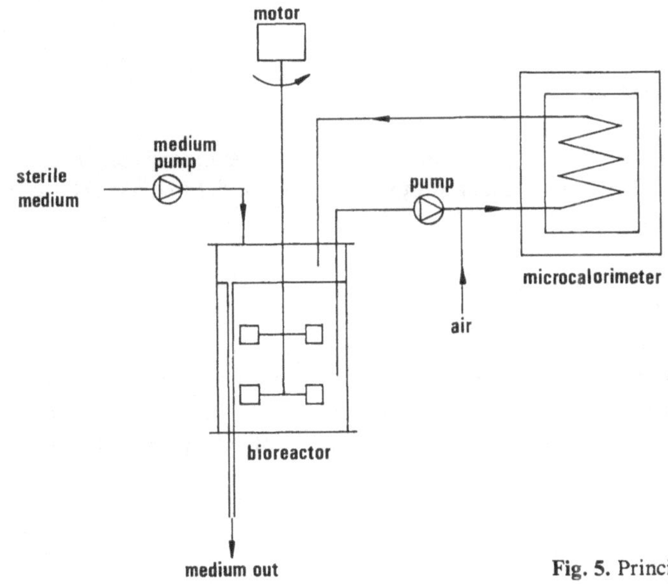

Fig. 5. Principles of flow microcalorimetry

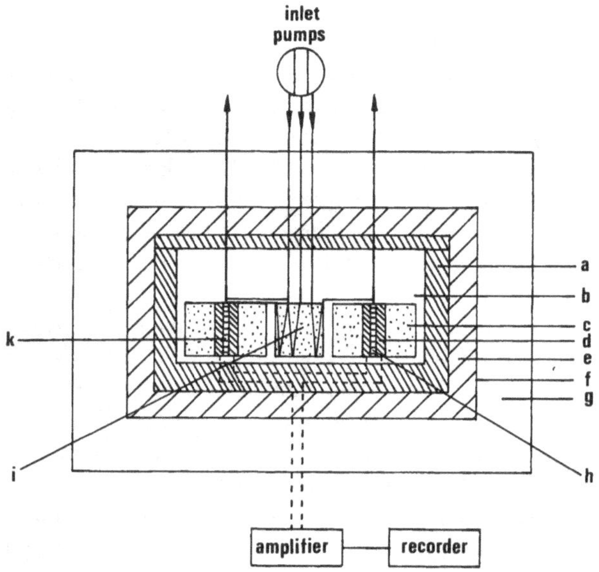

Fig. 6. Schematic representation of a flow microcalorimeter (*a*) main heat sink (*b*) air space (*c*) aluminium block (*d*) semiconductor thermopiles (*e*) styrofoam insulation (*f*) stainless steel container (*g*) thermostated water bath (*h*) flow-through cell (*i*) heatexchanger unit (*k*) mixing reaction cell. (Re-drawn with permission from [39])

is the need for care in the use of electrical calibration of the system due to heat losses in the sample flow [51]. Such problems make the reproducible quantitative interpretation of the calorimetric data obtained, under strictly defined culture conditions, difficult to achieve, even though the sensitivity of such devices is very high (2–3 mW L^{-1}).

2.3 Isothermal Reaction Calorimeters

These instruments are characterized by the use of a modified standard benchscale reactor in which the heat flux can be measured in situ. These are defined as "macro" or bench-scale calorimeters.

2.3.1 Dynamic Calorimetry

Cooney et al. [24] used a system based on the development of an energy balance around a carefully insulated 14-liter (10 L working volume) vessel, when the cooling system was temporarily inactivated. The increase in the culture temperature over a small temperature range enables the determination of the heat evolved by the biological process from Eq. (2)

$$K_P \frac{dT}{dt} = q_R + q_A - q_L - q_G , \tag{2}$$

where K_P is the total heat capacity of the reactor (J °C^{-1}), q_R is the heat generated by the bioreaction (W), q_L represents the heat losses (W), q_A is the heat generated through agitation (W) and q_G represents the heat lost to the gas-stream (W). The terms q_A, q_G and q_L are determined through electrical calibration of the system before inoculation.

After inoculation, the cooling system is switched off for periods of 5 to 9 min and the rate of temperature increase of the culture measured over a temperature change of approximately 1 °C.

This basic method [24] has been improved by a number of workers [27,52]. However, a detailed knowledge of each of the heat loss and gain terms on the energy balance is required. The value of q_G can be reduced by thermostating and pre-saturation of the airstream entering the bioreactor and by reducing heat losses by thermostating the top-plate.

Whilst this "macro" or bench-scale calorimetric method operates in situ, it suffers from the need for a relatively high biomass concentration (> 1–2 g L^{-1}) since the sensitivity is low, and from the discontinuous nature of the measurement. Therefore, it is difficult to measure sudden changes in the heat generation signal such as those obtained during the late exponential phase of batch growth. Since the system is not rigorously isothermal, the temperature changes involved in the measurement may result in significant changes in the growth kinetics of the culture.

2.3.2 Continuous Calorimetry

In order to increase the sensitivity of the "dynamic method" and to permit continuous measurement of the heat generated during bench-scale bioreactions in situ, a group of workers [29,53] developed a so-called Continuous Calorimetry system (Fig. 7).

This system is composed of an insulated 14-liter standard laboratory bioreactor. Cooling water, of a precisely controlled temperature, is supplied through a cooling coil such that heat has to be supplied, through a variable output electrical heater, in order to maintain the temperature constant. The power supplied to the heater is

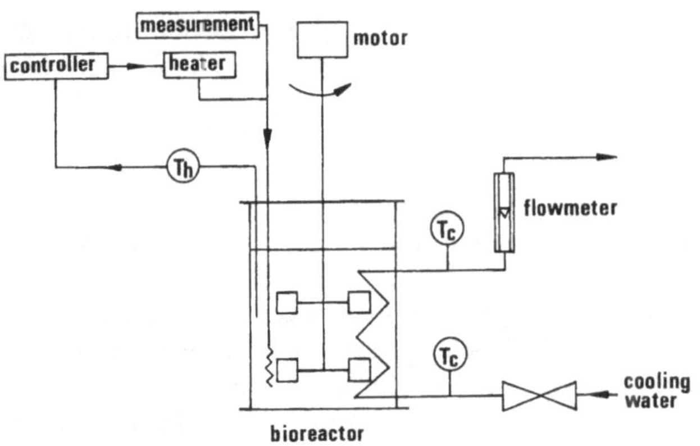

Fig. 7. Schematic diagram showing the principles of continuous calorimetry

thus inversely related to the heat generated by the culture. An energy balance around the bioreactor is still required in order to determine the heat generated by the culture (Eq. (3)).

$$q_R = q_H - q_C + q_A - q_L - q_G \tag{3}$$

where q_R, q_L, q_A, and q_G represent the heat generated by the culture, heat losses to the environment, heat generated through agitation and heat lost to the airstream (W), respectively. q_H represents the heat released by the electrical heater, while q_C is the constant cooling rate. As with dynamic calorimetry, the majority of these terms can be determined by calibration of the system before inoculation. Equations have been developed for correction of q_C for the response of the heater and the time constant of the system as a function of the flow of cooling liquid [29].

This in situ "macro"-calorimetric system requires an energy balance and consequently suffers from similar problems to those of dynamic calorimetry. However, the measurement is continuous and has a higher degree of sensitivity than the dynamic method [27].

2.3.3 Heat Flux Calorimetry

A relatively new in situ, isothermal, "macro" — or bench-scale calorimeter system (BSC-81) has been developed by Ciba-Geigy AG, Basel, Switzerland [54–56] and recently commercialized for chemical reactions by Mettler Instruments AG, Greifensee, Switzerland, as a "Reaction Calorimeter RC 1".

The BSC-81 heat flux calorimeter is composed of a 2-liter jacketed glass reactor, through the jacket of which is pumped a silicone oil at the rate of 2 L s^{-1}.

The oil circulatory system is composed of two parts (Fig. 8). In the first part oil is maintained at about 3 °C above the desired reactor temperature T_R. In the second part oil is cooled to a temperature of at least 5 °C below T_R. The two oil systems are connected by an electronic valve, the operation of which results in the mixing of the "hot" and "cold" oil such that the temperature of the oil passing through the jacket of the reactor T_J, is sufficient to maintain the reactor temperature, T_R, constant.

Measurement of T_R and T_J is continuously monitored and, at any instant, the heat exchanged between the contents of the reactor and the jacket oil (q_F) is given by Eq. (4).

$$q_F = UA(T_R - T_J) \tag{4}$$

where U is the global heat transfer coefficient (W K^{-1} m^{-2}) and A is the heat transfer surface (m^2).

If heat is generated by a reaction, the temperature of the jacket oil, T_J, is diminished in order to maintain T_R constant. The adjustments of T_J are effected by the automatic operation of the electronic mixing valve, which is controlled by an LSI 11 computer. This computer also mediates the continuous in situ measurement of the heat flux by evaluating Eq. (4) or by using a more sophisticated non-steady state model of the heat flux through the reactor wall.

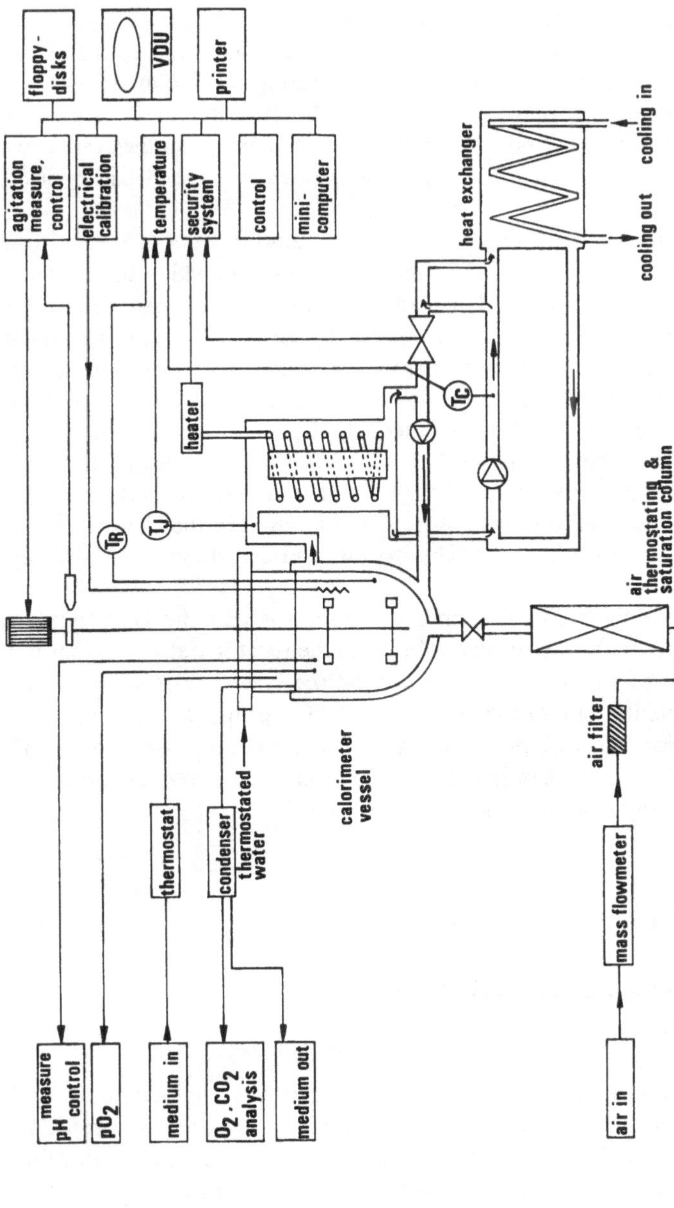

Fig. 8. Schematic diagram showing the principles of heat-flux calorimetry (BSC 81)

An electrical calibration heater installed in the vessel is used to determine the proportionality coefficient A · U as often as needed. It is thus possible to correct the measurement for changes of the heat transfer characteristics which may result from modifications of the viscosity, volume and wall growth.

This calorimetric system can be operated in isothermal, isoperibol or adiabatic modes, of which the isothermal mode is most suited for biological studies.

A more detailed explanation of the principle of the BSC-81, the adaptations necessary for use in biological studies, together with batch, aerobic studies involving a range of microorganisms grown on a range of substrates have been reported [4,30,31,57,58].

The calorimeter system is ideally suited to biological studies since the vessel has a design [30] comparable with a standard laboratory bioreactor in which the contents can be agitated, aerated, the pH controlled, and antifoam added. The in situ continuous measurements and temperature control are mediated by an LSI-11 computer which can be programmed to calibrate the system, through an electrical calibration heater, in order to correct for wall effects and viscosity changes which may result in changes in the value of the heat transfer coefficient, U. Furthermore, the system can be used for aerobic or anaerobic, batch, fed-batch or continuous culture studies.

The response time (50 s) and sensitivity ($\leqq 0.1$ W L^{-1}) are somewhat better than for other "macro"-calorimetric methods and, although less sensitive than many microcalorimetric methods (1–10 mW L^{-1}), can be operated under standard laboratory bioreactor conditions.

A major advantage of this continuous, in situ, method is that the heat signal is not affected by heat losses to the environment through the reactor walls and does not require the development of an energy balance around the system. Furthermore, the base-line signal represents the sum of the heat loss and gain terms, through agitation, aeration, nutrient addition, etc and, providing these effects are maintained constant, the heat generated by the culture can be determined from direct comparison of the measured heat signal at any instant to the base-line value.

3 Theory

3.1 Simultaneous Elemental and Enthalpy Balance for Cellular Growth Processes

If calorimetry is to play a more important role in quantitative engineering related studies and in bioreactor control, the meaning of microbial heat release must be understood and the quantitative relationship of the heat evolution rate with other relevant process variables, such as biomass concentration, growth rate, CO_2 evolution rate, and so on must be elucidated. Heat has often been considered as nonspecific information, which may account for some prejudices in the field of biocalorimetry. However, it has been shown in several papers [4,57,59,60] that specific, quantitative information on the above mentioned parameters may readily be deduced from heat-evolution measurements by analyzing them in terms of combined enthalpic and elemental balances of microbial growth. The stoichiometry of a general growth process giving rise to a maximum of one major product other than cell mass, CO_2, and water may be described in terms of a "chemical equation" as follows [4]:

$$CH_{s_1}O_{s_2}N_{s_3} + Y'_{O/S}O_2 + Y'_{N/S}NH_3 \rightarrow$$
$$Y'_{X/S}CH_{x_1}O_{x_2}N_{x_3} + Y'_{P/S}CH_{p_1}O_{p_2}N_{p_3} + Y'_{C/S}CO_2 + Y'_{W/S}H_2O \tag{5}$$

This equation has been formulated in terms of C-moles, which means that each chemical formula has been reduced to the basis of one carbon atom. In this notation, the letters appearing in the subscript stand for either substrate (S), biomass (X), product (P), O_2 (O), NH_3 (N), CO_2 (C) or H_2O (W), whereas the numbers designate the elements H (1), O (2), and N (3). The stoichiometric coefficients, or C-molar yields are defined as the ratio of two conversion rates r'_i (C-mol s^{-1} L^{-1}) of either substrates ($r'_i \leqq 0$) or products ($r'_i \geqq 0$):

$$Y'_{i/j} = \left|\frac{r'_i}{r'_j}\right| \tag{6}$$

They are related to conventional yield factors as follows:

$$Y'_{i/j} = \frac{(1 - A_i)}{(1 - A_j)} \cdot \frac{M'_j}{M'_i} \cdot Y_{i/j} \tag{7}$$

where A_i denotes the ash fraction of compound i ie, the fraction of elements other than C, H, O, N and M'_i the formula mass of one C-mole of i.

The relationship of the heat generation rate q (W L^{-1}) to some conversion rate may be expressed by a similar ratio:

$$Y'_{Q/i} = \left|\frac{q}{r'_i}\right| \tag{8}$$

The proportionality coefficients defined by Eq. (8) can be regarded as "heat yields" or "energetic yields". All yields are coupled through elemental and enthalpic balances. Although the six Y-coefficients appearing in Eq. (5) may vary in time, they must satisfy the four elemental balances for C, H. O, N at all times. This means that only 2 of the six yields are truly independent, and that it is possible to calculate all of them by solving the elemental balances, provided that two Y'-values are known in some independent way, e.g. through experimental determination.

Assuming that the biomass yield $Y'_{X/S}$ and the product yield $Y'_{P/S}$ are known, the solution of the balances yields [4]:

$$Y'_{O/S} = 1/4\,\gamma_S - 1/4\,\gamma_X Y'_{X/S} - 1/4\,\gamma_P Y'_{P/S} \tag{9}$$

$$Y'_{N/S} = s_3 + x_3 Y'_{X/S} + p_3 Y'_{P/S} \tag{10}$$

$$Y'_{C/S} = 1 - Y'_{X/S} - Y'_{P/S} \tag{11}$$

$$Y'_{W/S} = \delta_S - \delta_X Y'_{X/S} - \delta_P Y'_{P/S} \tag{12}$$

The expressions γ_i and δ_i appearing in these equations are defined for any substance i of the generalized chemical formula $CH_{e_{i1}}O_{e_{i2}}N_{e_{i3}}$ as:

$$\gamma_i = 4 + e_{i1} - 2e_{i2} - 3e_{i3} \tag{13}$$

$$\delta = 1/2\, e_{i1} - 3/2\, e_{i3} \tag{14}$$

The usefulness of heat evolution measurements lies in the fact that heat yields defined by Eq. (8) are related to the other yields by an enthalpy balance:

$$\sum_i Y'_{i/s}\, \Delta H'_i = Y'_{Q/S} \tag{15}$$

In this equation, the stoichiometric coefficients referring to products are counted negatively, and those concerning the reactants positively. By substituting Eq. (10) for $Y'_{N/S}$ one can express, for instance, the amount of heat evolved per C-mole of substrate consumed ($Y'_{Q/S}$) as a function of the same two independently known yield coefficients in the same way as for all other yield factors:

$$\begin{aligned} Y'_{Q/S} = \Delta H'_S - s_3\, \Delta H'_N - Y'_{X/S}(\Delta H'_X - x_3\, \Delta H'_N) \\ - Y'_{P/S}(\Delta H' - p_3\, \Delta H'_N \end{aligned} \tag{16}$$

Measuring a minimum of two yields is therefore sufficient to determine the whole stoichiometry of the system including the heat yields, (the demonstration has been made here assuming that $Y'_{X/S}$ and $Y'_{P/S}$ would be measured and using $Y'_{Q/S}$ as an example for a heat yield, but as will be shown later, the equations could easily be rewritten in terms of two independently known yields other than $Y'_{X/S}$ and $Y'_{P/S}$ and using different heat yields).

Now, in cases where not even two, but only one of the stoichiometry coefficients can be determined by measurement, one might resort to a heat evolution measurement and determine the missing yield on the basis of an equation such as (16). Thus, the whole stoichiometry could be determined using, for instance, $Y'_{Q/S}$ and $Y'_{X/S}$.

In cases where two or more stoichiometry coefficients are measurable, the determination of a heat yield is still highly useful in that it permits checking the consistency of the data and also improving the accuracy of the estimations of the other yield measurements by applying the balance equations and appropriate statistical parameter estimation procedures. These possibilities actually represent some of the important benefits of measuring microbial heat dissipation rates quantitatively.

3.2 Elemental and Enthalpic Contents of Substrate and Products and the Role of the Degree of Reduction

In order to use the combined elemental and enthalpy balances set forth in Sect. 3.1, one needs the chemical formula for all compounds listed in Eq. (5) as well as data on their heats of combustion. While such data is readily available in reference tables for all chemically defined components, it appears difficult to estimate the enthalpic content of dried microbial biomass from the literature because of a considerable scatter of sometimes as much as 50%[20,21,61-65]. It may therefore be necessary to determine the heat of combustion of the dried biomass experimentally using combustion calorimetry. Unfortunately, such experiments are tedious and the preparation of representative samples requires great care[66], which may explain some of the scatter.

The "chemical formula" of dried biomass is easier to measure and the literature data on this subject is much better. An attractive alternative to combustion calorimetry is therefore the estimation of $\Delta H_X'$ on the basis of elemental composition.

Several mathematical models relating the heat of combustion of dried biomass to its elemental composition have recently been critically reviewed by Cordier et al.[66]. They were compared with each other both from a theoretical point of view and experimentally by performing combustion calorimetry and elemental analysis for a number of microbial strains. The results of some of the experimental determinations are summarized in Table 3.

Although these literature models differ widely in form, it was demonstrated that all of them boil down to the idea that the heat of combustion is proportional to the number of moles of oxygen required for combustion. This is hardly surprising in view of the fact that Thornton[67] had shown as early as 1917 that the proportionality coefficient between the heat of combustion and the number of moles of O_2 was very nearly the same for many organic chemicals, a finding which was later extended to an enormous number of compounds by Kharasch[68].

The number of moles of oxygen consumed during combustion of one C-mol of a compounds is found from the following stoichiometry:

$$\begin{aligned} &CH_{e_{i1}}O_{e_{i2}}N_{e_{i3}} + (1 + 1/4\, e_{i1} - 1/2\, e_{i2})\, O_2 \rightarrow \\ &CO_2 + 1/2\, e_{i1} H_2O + 1/2\, e_{i3} N_2 \end{aligned} \tag{17}$$

Kharasch unfortunately chose not to express the universal proportionality coefficient per mol O_2 but rather per equivalent of electrons transferred to oxygen during

Table 3. Elemental composition and heat of combustion of dry biomass (recalculated from Cordier et al.[66] $\Delta H_x'$ and M_x' are on a dry, ash-free basis)

Organism	Substrate	x_1	x_2	x_3	γ_x	M_x' g (C-mol)$^{-1}$	$\Delta H_x'$ calculated according to Eq. (19) kJ (C-mol)$^{-1}$	$\Delta H_x'$ measured kJ (C-mol)$^{-1}$
E. coli	Glucose	1.701	0.424	0.250	4.853	23.99	558.1	552.7
M. methylotrophus	Methanol	1.716	0.400	0.247	4.917	23.58	565.5	561.6
K. fragilis	Lactose	1.776	0.575	0.159	4.627	25.19	532.1	542.6
K. fragilis	Galactose	1.751	0.530	0.166	4.690	24.56	539.4	534.9
K. fragilis	Glucose	1.748	0.517	0.153	4.713	24.16	542.0	523.3

combustion. The number of transferred electrons, presently termed the "degree of reduction" γ_i^0, is obviously 4 times the number of moles of oxygen consumed:

$$\gamma_i^0 = 4 + e_{i1} - 2e_{i2} \tag{18}$$

Therefore

$$\frac{\Delta H'i}{\gamma_i^0} \approx \text{constant} = Q_0 \tag{19}$$

Equation (19) or a very similar one is the explicit or implicit basis for all models proposed to estimate the heat of combustion of dry biomass, although different models assign different values to the universal constant Q_0. In the comparative study of Cordier et al. [66], the value proposed by Giese [69] and also by Roels [70] of $Q_0 = 115$ kJ C-mol^{-1} eq electrons)$^{-1}$ gave the best match with the experimental results (see Table 3).

The reductance degree γ_i^0 defined in Eq. (18) strikingly resembles the function γ_i (Eq. 13) appearing in the solutions of the elemental balances. This stems from the fact that the overall stoichiometry of microbial growth can be broken down into three parallel "combustion" reactions of the following form:

$$CH_{e_{i1}}O_{e_{i2}}N_{e_{i3}} + 1/4(4 + e_{i1} - 2e_{i2} - 3e_{i3})\,O_2 \rightarrow$$
$$CO_2 + (1/2\,e_{i1} - 3/2\,e_{i3})\,H_2O + e_{i3}NH_3 \tag{20}$$

In order to obtain the overall stoichiometry Eq. (5), one writes Eq. (20) for the substrate, rewrites it in the inverse sense for the biomass and the product by multiplying it by, respectively, $Y'_{X/S}$ and $Y'_{P/S}$, and adds the three expressions. Comparison with Eq. (5) immediately yields Eqs. (9)–(12). Therefore, the function γ_i (Eq. (13) indicates four times the number of moles of O_2 consumed by the combustion described by Eq. (20). It can thus be considered to be the reduction degree of i defined for a combustion yielding the nitrogen in the form of NH_3 rather than N_2. This "generalized" reduction degree γ_i was introduced by Minkevịch and Eroshin [71] and by Erickson [73] and discussed in details by Roels [70]. The relationship between the two reduction degrees is found by comparison:

$$\gamma_i = \gamma_i^0 - 3e_{i3} \tag{21}$$

It is obvious that the enthalpy balances could be simplified if the enthalpy of combustion was defined according to Eq. (20) rather than Eq. (17). As an accordingly modified heat of combustion ΔH_i^* would assign an enthalpy of zero to NH_3, the need for an explicit term for ammonia in the enthalpy balances would be eliminated. Eq. (16) could thus be replaced by:

$$Y'_{Q/S} = \Delta H_S^* - Y'_{X/S}\,\Delta H_X^* - Y'_{P/S}\,\Delta H_P^* \tag{22}$$

By applying Hess' Law to Eqs. (17) and (20), the relationship between $\Delta H_i'$ and ΔH_i^* is found as

$$\Delta H_i^* = \Delta H_i' - e_{i3}\,\Delta H_N' \tag{23}$$

Substitution of the modified combustion enthalpies ΔH_i^* in Eq. (22) by Eq. (23) yields the balance in its conventional form, Eq. (16).

ΔH_i^* and γ_i have the same relation as $\Delta H_i'$ and γ_i^0. This is shown by dividing Eq. (23) by Eq. (21) and allowing for Eq. (19), which also holds true with less than 10% of error for ammonia, for which γ_i^0 is 3:

$$\frac{\Delta H_i^*}{\gamma_i} = \frac{Q_0\gamma_i^0 - e_{i3}Q_0 \times 3}{\gamma_i^0 - 3e_{i3}} = Q_0 \tag{24}$$

In the case of substrate and product, ΔH_i^* often reduces to $\Delta H_i'$ because they rarely contain nitrogen, which makes s_3 and p_3 zero.

3.3 Predicting Important Heat Yields

An obvious use of the microbial heat evolution data would be the indirect determination of biomass concentration. This application would require some knowledge of the amount of heat per unit biomass formed, ie. on $Y'_{Q/X}$. On the other hand, the question arises as to whether interesting information could be derived from a systematic measurement of this heat yield $Y'_{Q/X}$.

An expression for this heat yield is obtained by dividing Eq. (22) by $Y'_{X/S}$ and allowing for Eq. (8):

$$Y'_{Q/X} = \frac{\Delta H_S^* - Y'_{P/S}\,\Delta H_P^*}{Y'_{X/S}} - \Delta H_X^* \tag{25}$$

If neither substrate nor product contain nitrogen, this can be written explicitly as follows:

$$Y'_{Q/X} = \frac{\Delta H_S' - Y'_{P/S}\,\Delta H_P'}{Y'_{X/S}} - \Delta H_X' + x_3\,\Delta H_N' \tag{26}$$

In Eq. (25), the modified heats of combustion could be estimated using Eq. (24). The heat yield then becomes

$$Y'_{Q/X} = Q_0\left[\frac{\gamma_s - Y'_{P/S}\gamma_P}{Y'_{X/S}} - \gamma_x\right] \tag{27}$$

where Q_0 is about 115 kJ C-mol^{-1} (degree of reduction)$^{-1}$.

The amount of heat released per C-mol of biomass formed is thus expected to increase for microbial growth on more reduced substrates. Furthermore, more heat will be dissipated per unit biomass in less efficient growth processes, i.e., with decreasing $Y'_{X/S}$. The appearance of a bioproduct will probably lower $Y'_{X/S}$, thereby tending to increase the heat dissipation. A substantial amount of the chemical energy available in the substrate could, however, be conserved in the product, reflected by the term $Y'_{P/S}\gamma_P$. This may nor may not overcompensate the influence of $Y'_{X/S}$ such that the net effect of product generation on the heat yield could be positive or negative. At any rate, it should be possible to detect the appearance of a product by monitoring $Y'_{Q/X}$. As γ_x stays fairly constant from one strain to another (Table 3), it will not be possible to distinguish between different organisms on the basis of $Y'_{Q/X}$ as long as the biomass and product yields are similar.

The heat dissipated per mole of oxygen consumed, designated by $Y'_{Q/O}$, can be predicted by dividing Eq. (22) by Eq. (9):

$$Y'_{Q/O} = \frac{\Delta H_S^* - Y'_{X/S}\,\Delta H_X^* - Y'_{P/S}\,\Delta H_P^*}{1/4(\gamma_S - Y'_{X/S}\,\gamma_X - Y'_{P/S}\,\gamma_P)} \tag{28}$$

Eliminating ΔH_i^* by means of Eq. (24) yields:

$$Y'_{Q/O} = 4\,Q_0 = 460\ \text{kJ mol}^{-1} \tag{29}$$

The fact that the amount of heat released per mole of oxygen consumed should be nearly the same for all strongly aerobic growth processes regardless of microbial strain, nature of substrate or product, has been discussed by many authors [57,70,72,73]. However, it must be borne in mind that this result becomes invalid in weakly aerobic or anaerobic processes, as it would predict a zero heat effect. In reality, the various deviations from the correlations expressed in Eqs. (19) and (24) will always ensure a positive heat effect even in the absence of any oxygen consumption.

The ratio of heat to CO_2 evolved could be useful because both evolution rates can be measured on-line in industrial bioreactors. This ratio is related to $Y'_{X/S}$ and $Y'_{P/S}$ as follows:

$$Y'_{Q/C} = Q_0\,\frac{\gamma_S - Y'_{X/S}\,\gamma_X - Y'_{P/S}\,\gamma_P}{1 - Y'_{X/S} - Y'_{P/S}} \tag{30}$$

Using these procedures, a number of other heat yields could be formulated and related to $Y'_{X/S}$ and $Y'_{P/S}$ or to another pair of C-molar yields defined by Eq. (6). The choice of the set of yield coefficients will mainly depend on what rate information other than exothermicity can be conveniently measured and on the specific information one wishes to deduce from the on-line measurement of microbial heat production. On the other hand, the choice of the set of yields cannot be based solely on this criterion, as singularities have been shown to exist for certain yield factors by Stephanopoulos and San [74-77]. For a detailed discussion of these problems the reader is referred to their series of papers.

3.4 Modelling Special Cases

Important special cases arising in biotechnology include purely aerobic, respiratory growth processes, purely anaerobic processes and defined mixtures of the two such as observed in many aerobic yeast growth processes. The stoichiometry of the former two types of microbial growth is simpler than the general stoichiometry described in Eq. (5).

Pure aerobic growth processes involve respiration as a major metabolism and are characterized by the absence of an important product other than biomass, CO_2 and water. Their stoichiometry should therefore be described as follows:

$$\begin{aligned} &CH_{S_1}O_{S_2}N_{S_3} + (Y'_{O/S})_R\, O_2 + (Y'_{N/S})_R\, NH_3 \rightarrow \\ &(Y'_{X/S})_R CH_{X_1}O_{X_2}N_{X_3} + (Y'_{C/S})_R CO_2 + (Y'_{W/S})_R H_2O \end{aligned} \tag{31}$$

The 5 Y'-coefficients of this equation are coupled by 4 elemental balances, which leaves only one truly independent yield. All stoichiometric coefficients can be expressed as a function of one single yield, which also means that the stoichiometry can be completely determined based on a single measured yield. It would therefore be possible to monitor the biomass yield indirectly and on-line by monitoring, for instance, the heat and CO_2 evolution rates and by means of Eq. (30) provided that γ_X and γ_S are sufficiently different. The necessary expressions are obtained from the previously discussed equations by setting $Y'_{P/S} = 0$ and are reported in Table 4.

Pure anaerobic growth is assumed to involve catabolic processes giving rise to substrate-level phosphorylation only and are characterized by their independence of an external electron acceptor. The respective stoichiometry might therefore be described as:

$$\begin{aligned} &CH_{S_1}O_{S_2}N_{S_3} + (Y'_{N/S})_F\, NH_3 \rightarrow \\ &(Y'_{X/S})_F\, CH_{X_1}O_{X_2}N_{X_3} + (Y'_{P/S})_F\, CH_{P_1}O_{P_2}N_{P_3} + (Y'_{C/S})_F\, CO_2 \\ &+ (Y'_{W/S})_F\, H_2O \end{aligned} \tag{32}$$

Here again, the stoichiometry is entirely fixed if one single yield coefficient is known. The stoichiometric coefficients in Eq. (32) have been expressed as a function of $(Y'_{X/S})_F$ by several authors [59, 70]. The heat yields shown in Table 4 have been determined using these expressions for the stoichiometric coefficients.

Although mixed metabolism between pure aerobic and pure anaerobic growth is, in principle, covered by the general equation of Sects. 3.1 and 3.3, it is possible in more defined cases of microbial growth to eliminate one of the two independent yield factors ($Y'_{X/S}$ and $Y'_{P/S}$) by making use of the fact that they are obviously interrelated. von Stockar and Birou [59] developed a theoretical model for this case by describing all phenomena as the result of two "parallel chemical" reactions, given by Eqs. (31) and (32). They then evaluated the stoichiometric coefficients appearing in Eq. (5) and the heat yields as a function of the relative "mix" of respiratory and anaerobic metabolism by utilizing the calculation procedures usually applied in

Table 4. Heat yields for various special cases

1. Pure Respiration, no Fermentation Product

$$Y'_{Q/S} = Q_0[\gamma_s - \gamma_x(Y'_{X/S})_R] \tag{34}$$

$$Y'_{Q/X} = Q_0\left[\frac{\gamma_s}{(Y'_{X/S})_R} - \gamma_x\right] \tag{35}$$

$$Y'_{Q/O} = 4Q_0 \tag{36}$$

$$Y'_{Q/C} = Q_0 \frac{\gamma_s - \gamma_x(Y'_{X/S})_R}{1 - (Y'_{X/S})_R} \tag{37}$$

2. Anaerobic Growth by Fermentation

$$Y'_{Q/S} = \left(\Delta H^*_S - \frac{\gamma_S}{\gamma_P}\Delta H^*_P\right) - \left(\Delta H^*_X - \frac{\gamma_X}{\gamma_P}\Delta H^*_P\right)(Y'_{X/S})_F \tag{38}$$

$$Y'_{Q/X} = \frac{\Delta H^*_S - \frac{\gamma_S}{\gamma_P}\Delta H^*_P}{(Y'_{X/S})_F} - \left(\Delta H^*_X - \frac{\gamma_X}{\gamma_P}\Delta H^*_P\right) \tag{39}$$

$$Y'_{Q/C} = \frac{\left(\Delta H^*_S - \frac{\gamma_S}{\gamma_P}\Delta H^*_P\right) - \left(\Delta H^*_X - \frac{\gamma_X}{\gamma_P}\Delta H^*_P\right)(Y'_{X/S})_F}{\left(1 - \frac{\gamma_S}{\gamma_P}\right) - \left(1 - \frac{\gamma_S}{\gamma_P}\right)(Y'_{X/S})_F} \tag{40}$$

3. Mixed Catabolism

$$Y'_{Q/S} = \Delta H^*_S - \Delta H^*_X(Y'_{X/S})_R\,\Omega - \left[\frac{\gamma_S}{\gamma_P}\Delta H^*_P + \left(\Delta H^*_X - \frac{\gamma_X}{\gamma_P}\Delta H^*_P\right)(Y'_{X/S})_F\right](1 - \Omega) \tag{41}$$

$$Y'_{Q/X} = \frac{\Delta H^*_S - \left[\frac{\gamma_S}{\gamma_P} - \frac{\gamma_X}{\gamma_P}(Y'_{X/S})_F\right]\Delta H^*_P(1 - \Omega)}{(Y'_{X/S})_R\,\Omega + (Y'_{X/S})_F\,(1 - \Omega)} - \Delta H^*_X \tag{42}$$

$$Y'_{Q/O} = \frac{\Delta H^*_S - [(Y'_{X/S})_R\,\Omega + (Y'_{X/S})_F\,(1 - \Omega)]\,\Delta H^*_X - \left[\frac{\gamma_S}{\gamma_P} - \frac{\gamma_X}{\gamma_P}(Y'_{X/S})_F\right]\Delta H^*_P(1 - \Omega)}{\frac{1}{4}[\gamma_S - \gamma_X(Y'_{X/S})_R]\,\Omega} \tag{43}$$

chemistry for parallel reactions. The relative mix of catabolism was quantified by an "aerobicity parameter" Ω:

$$\Omega = \frac{4\,Y'_{O/S}}{\gamma_S - \gamma_X(Y'_{X/S})_R} \tag{33}$$

The aerobicity Ω describes the oxygen consumption of the culture relative to the substrate consumption, but is normalized by the maximum value for pure respiration. Thus, Ω varies from 1 for purely respiratory growth to zero for pure anaerobic growth. The model also contains two parameters, $(Y'_{X/S})_R$ and $(Y'_{X/S})_F$, which can be determined by independent growth experiments under completely aerobic and completely anaerobic conditions respectively.

The results for the most important heat yields are summarized in Table 4. As can be easily verified, these equations reduce to the solution for pure respiratory growth for $\Omega = 1$ and if Eq. (24) is substituted for ΔH_i^*. On the other hand, the equations for anaerobic growth result for $\Omega = 0$.

By a careful analysis of the derivative of Eq. (42) in Table 4 with respect to Ω, von Stockar and Birou [59] predicted that the amount of heat generated per unit biomass formed must be larger for respiratory than for fermentative growth in yeast. As may be seen from Eq. (42), it decreases continuously from the maximum value for respiration to the anaerobic value as the catabolism becomes more and more repressed. It was also concluded that the same would be expected for most facultative anaerobes. Exceptions are conceivable in cases with unusually high values of $(Y'_{X/S})_R$, very low values of $(Y'_{X/S})_F$ and low enthalpic content of the bioproduct.

The proportionality coefficient between heat evolution and oxygen uptake (Eq. (43)) reduces to a value of 4 Q_0 for $\Omega = 1$ and tends to infinity for $\Omega = 0$.

Equations (41)–(43) in Table 4 could be used to evaluate the "aerobicity parameter" Ω from a measured heat yield. As Ω determines the whole stoichiometry, a heat yield measurement could serve to determine all stoichiometric coefficients in the overall Eq. (5), including biomass yield. This again stresses the fact that heat dissipation measurements could, if utilized in an appropriate manner, constitute a powerful on-line "probe" for monitoring the state and the metabolism of the culture.

4 Applications

4.1 Measurement of Microbial Growth

Cultivation of microorganisms on defined minimal and complex media results in thermograms which have a distinct shape that can be used for the identification of specific strains [78–81]. However, it is clear from Sect. 3 that the total heat liberated will depend on the nature of the carbon and energy source, particularly the degree of reduction, the biomass yield coefficient and physical conditions such as temperature, pH and the aerobicity of the culture, as well as the specific chemical composition of the medium. For these reasons, the use of calorimetry for the "finger-printing" of strains must be undertaken under strictly controlled conditions [82, 83].

Inoculation of a suitable growth supporting medium results in a growth curve characteristic of batch cultures in which the biomass concentration increases exponentially until some component of the medium becomes limiting (Fig. 9a and b) [30]. The heat generation rate, q, follows the growth curve until the end of the exponential growth phase. What follows depends upon the nature of the limitation. Thus in carbon- and energy-limited cultures, the heat signal falls rapidly to a basal value defined by the endogenous metabolism of the cells (Fig. 9b). With cultures limited by the nitrogen source, the heat signal falls rapidly to a value considerably higher than that obtained for endogenous metabolism alone. Heat continues to be released at a constant rate until the carbon and energy source are exhausted (Fig. 10a and b). This latter

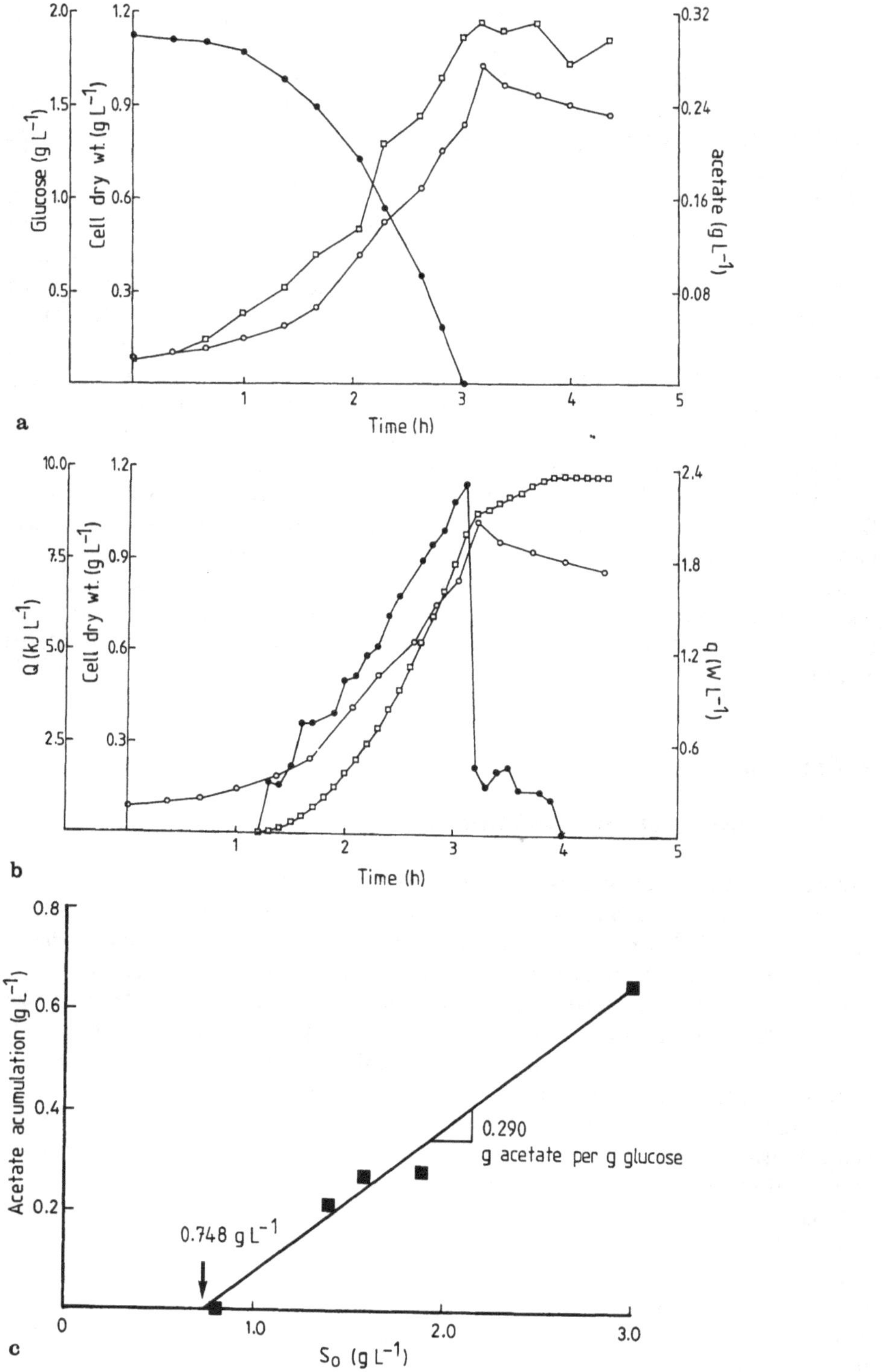
Glucose (g L−1)
Cell dry wt. (g L−1)
acetate (g L−1)
Time (h)
a
Q (kJ L−1)
Cell dry wt. (g L−1)
q (W L−1)
Time (h)
b
Acetate acumulation (g L−1)
0.290
g acetate per g glucose
0.748 g L−1
So (g L−1)
c

Fig. 9a–c

example represents a case of uncoupled growth in which carbon and energy source is metabolized, and heat released, for no net increase in biomass concentration [30].

Similar results have been obtained for Mg^{2+} limited batch cultures of the yeast *Kluyveromyces fragilis* grown on a minimal salts medium containing lactose as the

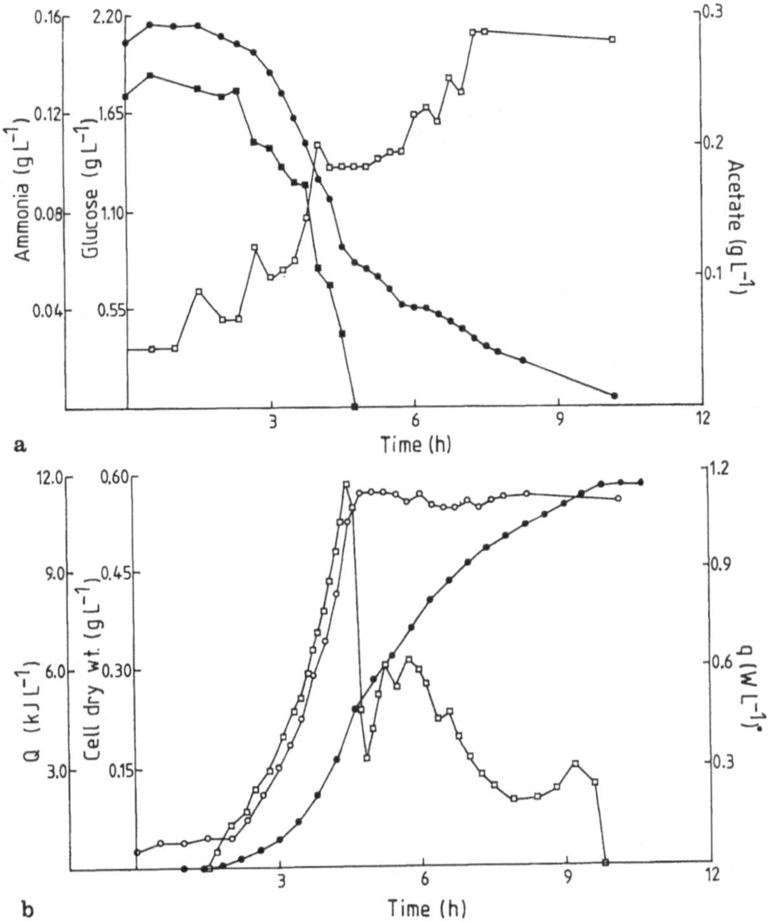

Fig. 10a. Growth of *E. coli* in a nitrogen-limited minimal salts medium. Symbols; (■) NH_4^+ concentration; (●) glucose concentration; (□) acetate concentration. (Re-drawn with permission from [30]) **b.** Calorimetric data for the growth of *E. coli* as Fig. 10a. Symbols: (□) heat evolution rate q; (●) total heat evolved Q; (○) cell dry weight. (Re-drawn with permission from [30])

◀ **Fig. 9.** Growth of *E. coli* in a glucose-limited minimal salts medium Symbols. (○) cell dry weight; (●) glucose concentration; (□) acetate concentration. (Re-drawn with permission from [30]. **b.** Calorimetric data for growth of *E. coli* as **a.** Symbols: (●) heat generation rate, q; (□) total heat evolved, Q; (○) cell dry weight. (Re-drawn with permission from [30]). **c.** Relationship between total acetate formed and the initial glucose concentration for *E. coli* grown as in **a.** (Re-drawn with permission from [30])

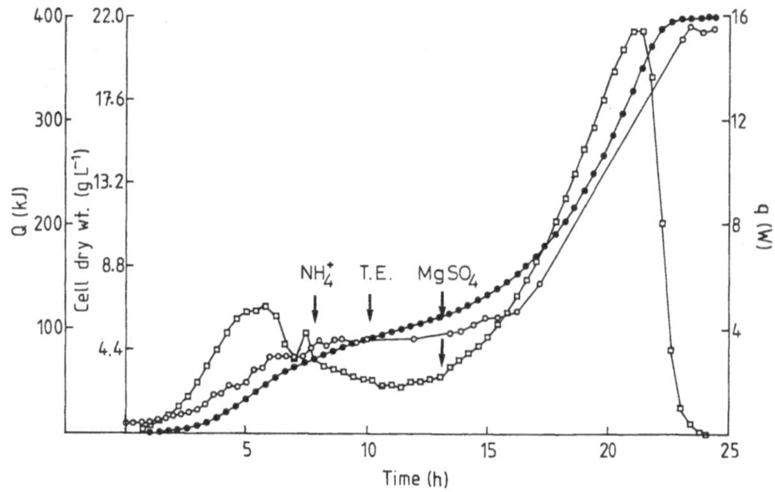

Fig. 11. Growth of *K. fragilis* on minimal salts medium containing lactose for detection of Mg^{2+}-limitation. Symbols: (○) cell dry weight; (□) heat evolution rate; (●) total heat evolved. Addition of NH_4^+ (as NH_4Cl), trace elements (T.E.) and; Mg^{+2} (as Mg Cl_2), respectively. Re-drawn with permission from Enzyme & Microbial Technology by permission of the publishers. Butterworth & Co. (Publishers) Ltd. © [4)]

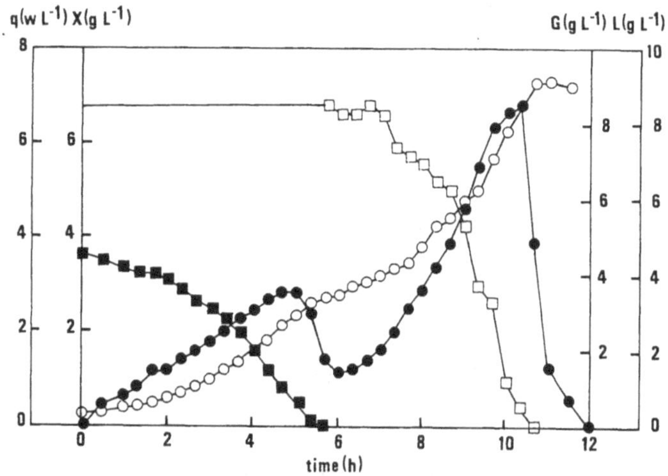

Fig. 12. Diauxic growth of *K. fragilis* on a minimal salts medium containing a mixture of glucose (5 g L^{-1}) and lactose (10 g L^{-1}). Symbols: (○) cell dry weight, X; (■) glucose concentration, G; (●) heat evolution rate, q; (□) lactose concentration, L [58)]

limiting substrate [4)] and by other workers [47, 84, 85)]. Figure 11 demonstrates that limiting factors can conveniently be identified by pulsing possible candidates into the culture and by observing the response in heat dissipation. Clearly calorimetry has a potential for medium optimization.

Growth of microorganisms in defined or complex media containing more than one source of carbon and energy is frequently characterized by diauxic growth patterns, in which one substrate is used preferentially. An example is the growth of *Kluyveromyces fragilis* on a defined medium containing a mixture of glucose and lactose [58]. Glucose was used preferentially for growth (Fig. 12). Lactose was only used after the glucose in the medium was exhausted and after a period of induction of the synthesis of the enzymes necessary for the transport and hydrolysis of lactose. Similar results have been reported for cultures of *E. coli* growing on an equimolar mixture of glucose and lactose [28]. In each case, the rate of heat generation closely followed the biomass curve for each of the growth phases.

4.2 Biomass Yield

A considerable amount of evidence has been published which shows that the amount of heat evolved by a microbial culture is directly related to growth and can be represented by the yield coefficient Y_{kJ} or $Y_{Q/X}$ (kJ heat evolved per g cell dry weight formed). Forrest et al. [86] were one of the first to show that the total heat evolved by an anaerobic glucose-limited batch culture of *Streptomyces faecalis* was directly related to the concentration of biomass formed. Since this time similar relationships have been shown for a wide range of microorganisms grown aerobically and anaerobically on a wide range of substrates [4,25,30,47,52,57,58,86–88].

The heat yield coefficient $Y_{Q/X}$ (kJ heat evolved per g cell dry weight formed) can be determined from the total heat evolved by the culture (kJ L^{-1}) as a function of the biomass concentration (g L^{-1}) or from the rate of heat evolution by the culture (W L^{-1}) as a function of the biomass production rate (g L^{-1} h^{-1}). Figures 13a and b show the values obtained using both methods for an aerobic batch culture of the yeast *Kluyveromyces fragilis* grown on a glucose-limited minimal salts medium. When determining $Y_{Q/X}$ from rates, the specific growth rate μ is generally assumed to be constant during the exponential growth phase and the rate of biomass production calculated as μX. Both methods yield excellent correlations although the actual values for $Y_{Q/X}$ obtained were slightly different. This difference is mainly due to the difficulty in determining accurately the value of the specific growth rate μ (h^{-1}) during all stages of batch growth. Thus, at low biomass concentrations, in the period immediately after inoculation, the rate of heat evolution is very small and close to the limits of calorimetric detection, whereas at the end of the exponential growth phase, q (the heat evolution rate, W L^{-1}) and μ vary as a function of the limiting substrate concentration, which makes measurement of μ difficult. Scatter and errors in the heat evolution rate measurements are effectively smoothed by integration of the signal. Clearly continuous culture calorimetric experiments, in which the growth rate can be maintained constant, should prove invaluable for the determination of $Y_{Q/X}$ (see Sect. 4.3).

As shown by Eq. (25), and Table 4, the biomass yield coefficient has a pronounced effect on the value of $Y_{Q/X}$. The effect of $Y_{X/S}$ on $Y_{Q/X}$ could be shown from the experiment depicted in Fig. 14a [98] in which *E. coli* was grown on a minimal salts medium in an aerobic batch culture. After approximately 4 h, growth factors were added in the form of yeast extract. Upon addition of yeast extract to the culture, the heat evolution rate rapidly increased and growth was visibly stimulated. Figure

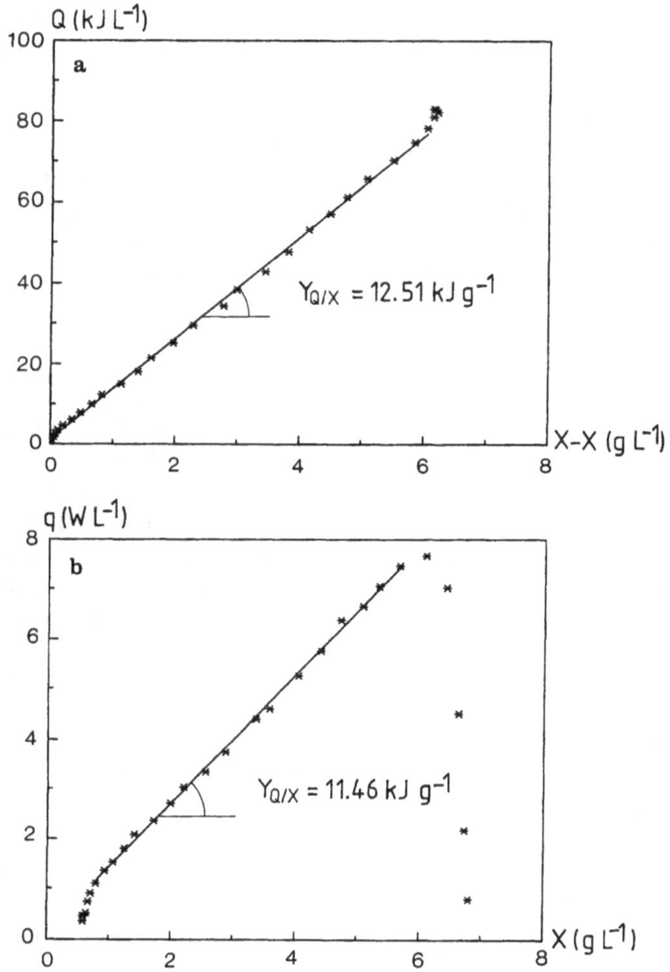

Fig. 13a. Determination of $Y_{Q/X}$ for *K. fragilis* grown aerobically on a glucose-limited minimal salts medium from measurements of the total heat evolved and biomass formed (Reprinted with permission from [57]). **b.** Determination of $Y_{Q/X}$ for *K. fragilis* grown as in Fig. 13a from measurements of the heat evolution rate and total biomass formed. Instead of plotting q vs μX, q was plotted as a function of X. The slope was divided by the maximum growth rate ($\mu = 0.404\ h^{-1}$) to evaluate $Y_{Q/X}$ (Reprinted with permission from [57]).

14b and c shows that not only the growth rate but also the biomass yield increased after addition of the yeast extract. The first growth phase was characterized by a high substrate consumption per unit biomass formed, and thus by low value of $Y_{X/S}$. The second phase was characterized by a lower substrate consumption and thus a higher value for $Y_{X/S}$ (Fig. 14c). Figure 14d shows how the total heat evolved varied as function of biomass formed, exhibiting three distinct phases. The first phase corresponds to a high level of heat dissipation per unit biomass ($Y_{Q/X}$) and is clearly influenced by the low value of $Y_{X/S}$. The second phase corresponds to a lower level

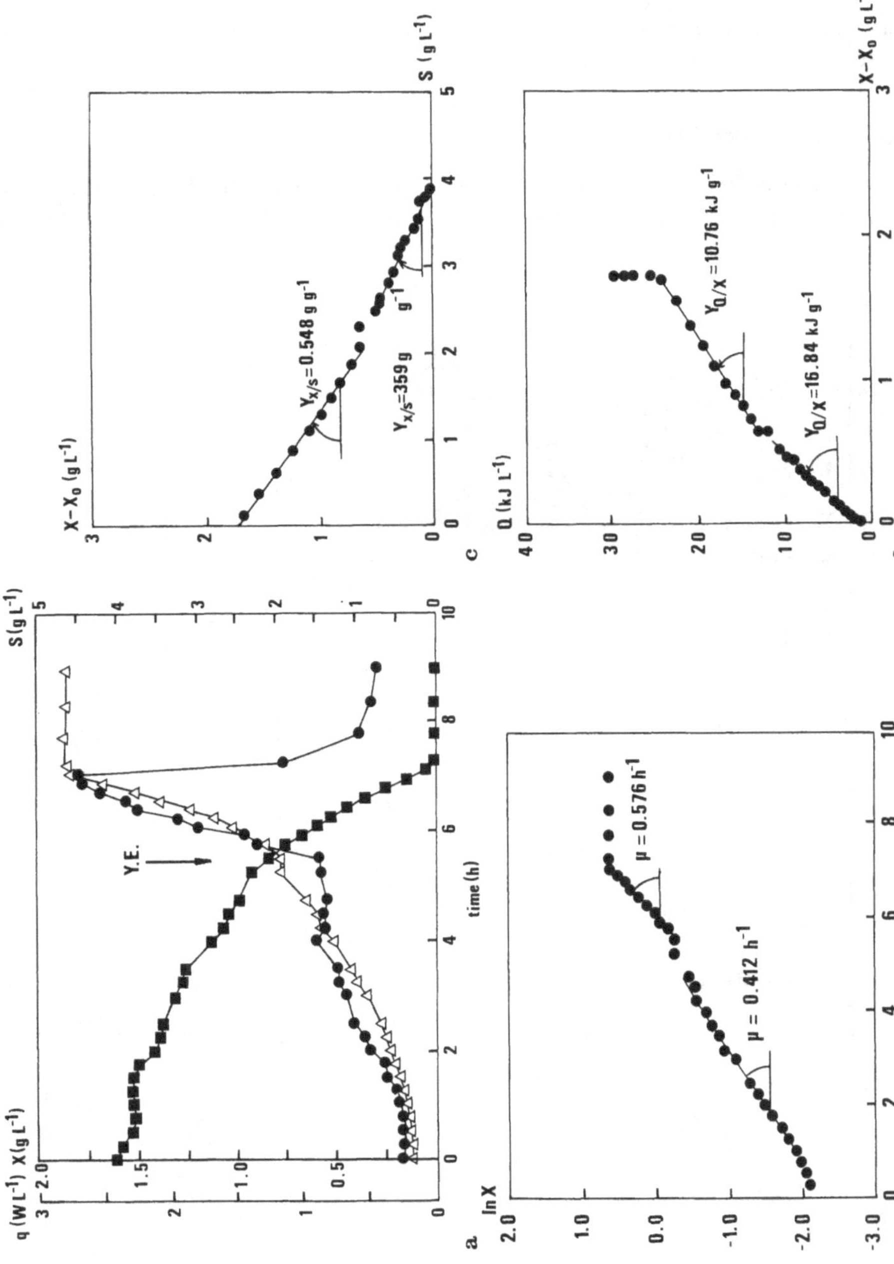

Fig. 14a. Thermogram of an aerobic batch culture of *E. coli* grown on glucose and the effect of yeast extract (YE) addition to the medium. Symbols: (△) cell dry weight, X; (●) heat evolution rate, q; (■) glucose concentration, S. (Re-drawn with permission from [98]). **b.** Effect of yeast extract addition on the growth rate of *E. coli*, Conditions as for **a** [98]. **c.** Effect of yeast extract addition on the biomass yield coefficient $Y_{X/S}$: $X-X_0$, cell dry weight formed: S, residual glucose concentration. Conditions as for **a** [98]. **d.** Effect of yeast extract addition on the specific heat yield coefficient $Y_{Q/X}$: Q, total heat evolved; $X-X_0$ cell dry weight formed. Conditions as for **a** [98]

of heat dissipation ($Y_{Q/X}$) resulting from a higher biomass yield ($Y_{X/S}$). The third phase in Fig. 14d is due to the non-growth-related oxidation of a by-product (acetate). Consequently, $Y_{X/S}$ is zero and $Y_{Q/X}$ tends to infinity.

4.3 Maintenance Energy

Calorimetric determinations at zero growth, provided they can be experimentally realized, afford a direct observation of the energy which must be dissipated by the

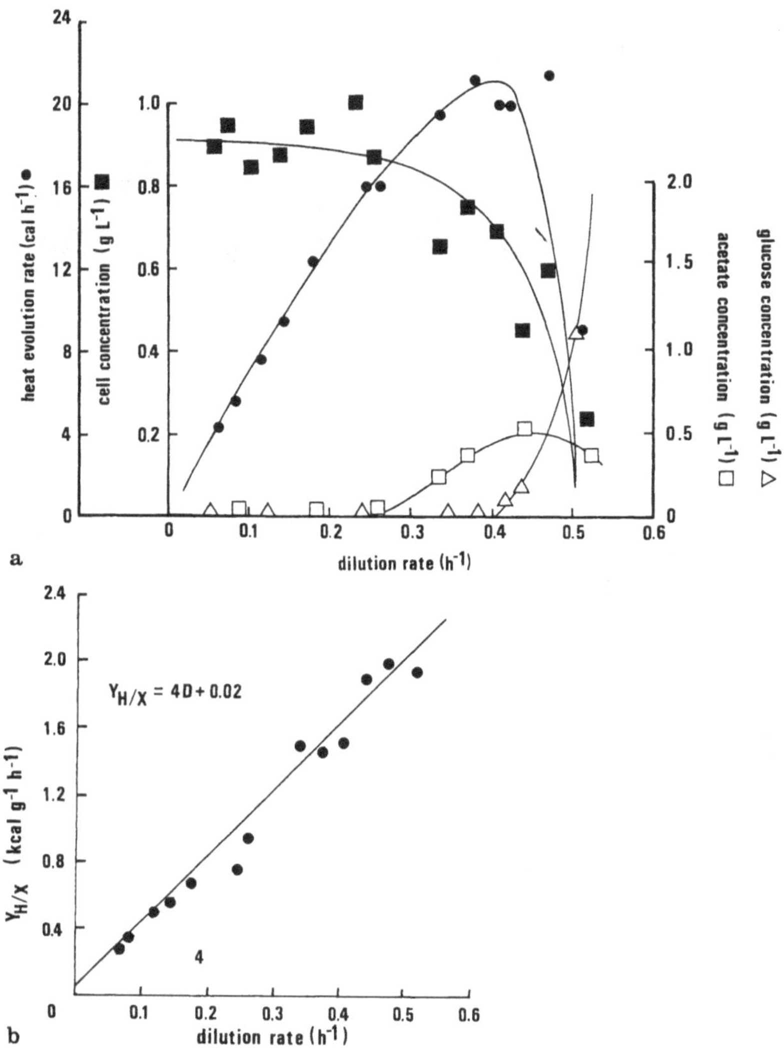

Fig. 15a. Heat evolution rate as a function of dilution rate for an aerobic glucose-limited chemostat culture of *E. coli* k-12 (Re-drawn with permission from [21]). **b.** Specific heat evolution rate (kcal per g cells per h) as a function of dilution rate for *E. coli* as predicted from Eq. (46). (Re-drawn with permission from [21])

culture for maintenance purposes. However reports of such measurements are scant and the results vary considerably. Estimates from batch experiments were reported by several authors and range from 0.1 to 0.2 W per g of dry biomass [21,29,89,90]. However, such estimates are unreliable and in some cases may have been affected by enthalpy balance errors due to undetected byproducts. Some attempts to determine the maintenance energy dissipation from continuous cultures have been reported [3,21,91]. In such experiments, the rate of heat dissipation is measured as a function of dilution rate and then extrapolated to zero by assuming that the heat released by the microbial culture can be broken down into heat associated with growth and with maintenance:

$$q = \frac{r_x}{Y^{max}_{X/Q}} + m_Q X \tag{44}$$

The observed heat yield $Y_{Q/X}$ is given by q/r_x. Also, as μX may be substituted for r_x, the observed heat yield should vary with specific growth rate as follows:

$$Y_{Q/X} = Y^{min}_{Q/X} + \frac{m_Q}{\mu} \tag{45}$$

The maintenance energy dissipation can thus be determined in principle as the slope of a plot of $Y_{Q/X}$ vs $1/D$. In practice, less scatter is obtained when using the following alternative form of Eq. (44) as the basis of a correlation.

$$\frac{q}{X} = Y^{min}_{Q/X} D + m_Q \tag{46}$$

Figure 15a shows the power released by a continuous culture of *E. coli* k-12 [21] as a function of dilution rate and Fig. 15b depicts a correlation according to Eq. (46). Ishikawa and Shoda [21] report that the intercept at the origin yielded a maintenance energy dissipation of 0.023 W g^{-1}.

Table 5 compares the results reported by three teams of authors having worked with chemostat cultures. The discrepancies could reflect differences from one micro-organism to another but might also be due to the fact that three different experimental techniques were used.

Table 5. Maintenance energy dissipation

Author	Year	Refs.	Experimental technique	Organism	m_Q W g^{-1}
Brettel et al.	1981	[91]	Flow calorimeter coupled to laboratory chemostat	*S. cerevisiae*	
				— defined medium	0.10
				— complex medium	0.05
Ishikawa and Shoda	1983	[21]	Microcalorimeter transformed into chemostat	*E. coli* k 12	0.023
Birou and von Stockar	1989	[3]	Bench scale calorimeter used as a chemostat	*K. fragilis*	0.025

4.4 Nature of Substrate and Growth Efficiency

The nature of the carbon and energy substrate has a profound effect on the heat released by a microbial culture. The amount of heat generated per unit biomass formed ($Y'_{Q/X}$) depends on the enthalpy content of the substrate as shown by Eq. (25) or can be related to its degree of reduction (γ_s) as shown by Eq. (27):

$$Y'_{Q/X} = Q_0 \left[\frac{\gamma_s - Y'_{P/S}\gamma_P}{Y'_{X/S}} - \gamma_x \right] \tag{27}$$

A study of microbial heat generation using a range of substrates with widely different degrees of reduction has recently been published by Birou et al. [57]. Their experiments were conducted under strongly aerobic conditions in order to suppress the formation of products as much as possible. The results of these experiments (Table 6) reflect the influence of γ_s on the heat yield and may be compared with the influence predicted by Eq. (27).

The heat yield $Y_{Q/X}$ clearly ranges from about 10 to 15 kJ g^{-1} cell dry weight when biomass is grown on substrates with an average degree of reduction, such as hexoses ($\gamma_s = 4$). As predicted by Eq. (27) much higher values of 16–26 kJ g^{-1} for $Y_{Q/X}$ are obtained with more reduced substrates such as ethanol, methanol, and hexadecane. Conversely, growth on more oxidized substrates such as citrate ($\gamma_s = 3$) and succinate ($\gamma_s = 3.5$) should yield lower value (<11 kJ g^{-1}) of $Y_{Q/X}$. However, the values obtained stay roughly in the same range as those observed with hexoses.

This apparent discrepancy is due to lower biomass yields $Y_{Q/X}$. The heat yield depends in an inversely proportional manner on $Y'_{X/S}$ as shown by Eq. (27) and as

Table 6. Heat yields as a function of reductance degree of substrate γ_s

Organism	Substrate	γ_s [–]	$Y_{X/S}$ [g g^{-1}]	$Y_{Q/X}$ [kJ g^{-1}]	$Y_{Q/O}$ [kJ mol^{-1}]
C. lipolytica	citrate	3	0.290	14.99	423
C. lipolytica	succinate	3.5	0.344	16.76	492
K. fragilis	Glucose (1)	4	0.520	12.51	447
*K. fragilis**	Glucose (2)	4	0.465	12.51	456
*K. fragilis**	Glucose (3)	4	0.388	15.16	—
*C. pseudotropicalis**	Glucose	4	0.465	13.50	408
C. utilis	Glucose	4	0.481	11.66	421
*E. coli**	Glucose	4	0.416	8.70	—
*Ent. cloacae**	Glucose	4	0.345	9.35	—
K. fragilis	Galactose	4	0.494	13.63	418
K. fragilis	Lactose	—	0.485	14.58	495
C. utilis	Acetate	4	0.406	17.78	385
C. utilis	Glycerol	4.67	0.562	10.79	474
C. boidinii	Ethanol	6	0.790	16.70	—
*C. utilis**	Ethanol	6	0.557	19.91	421
M. methylotrophus	Methanol	6	0.406	23.70	—
C. lipolytica	Hexadecane	6.13	0.986	25.63	437

demonstrated by experimental evidence discussed earlier. Indeed it has been pointed out by several authors that the C-molar biomass yield $Y'_{X/S}$ tends to decrease with decreasing γ_s below a critical value of about 4.5 because of the inherent energy limitation of oxidized substrates [60,70,92,93]. At γ_s values higher than the critical one, $Y'_{X/S}$ stays constant at about 0.6 C-mol/C-mol because of the inherent carbon limitation of such substrates:

$$Y'_{X/S} = a\,\gamma_s \quad \text{for} \quad \gamma_s < \gamma_s^c \tag{47a}$$

$$Y'_{X/S} = C \quad \text{for} \quad \gamma_s \geqq \gamma_s^c \tag{47b}$$

Roels [70] proposes the following typical values for the various constants in Eq. (47):

$$a = 0.13 \qquad C = 0.6 \qquad \gamma_s^c = 4.67$$

An equation predicting the observable heat yield as a function of γ_s should therefore incorporate this general behaviour of $Y'_{X/S}$ as a function of λ_s. Substituting Eq. (47) into Eq. (27) yields for the case of no product formation:

$$Y'_{Q/X} = Q_0\left[\frac{1}{a} - \gamma_x\right] \quad \text{for} \quad \gamma_s < \gamma_s^c \tag{48a}$$

$$Y'_{Q/X} = Q_0\left[\frac{\gamma_s}{C} - \gamma_x\right] \quad \text{for} \quad \gamma_s \geqq \gamma_s^c \tag{48b}$$

As shown in Fig. 16, the model expressed by Eq. (48) does indeed reflect the trends observed in Table 6 quite well. In conclusion, it appears possible on the basis of Eq. (48) to predict the heat yield roughly as a function of one single independent variable, namely the degree of reduction of the substrate, γ_s.

The heat yield and its relationship with the degree of reduction shown in Fig. 16 may be interpreted in terms of the efficiency of microbial growth energetics. As the heat yield is a direct measure of the chemical energy initially available in the substrate which has been dissipated as heat, it must be possible to deduce from it the fraction of the available energy not dissipated, i.e., retained in the biomass. In cases in which no products other than cell mass, CO_2 and water appear, this can be done by rearranging Eq. (26) as follows:

$$\frac{\Delta H'_S}{Y'_{X/S}} + x_3\,\Delta H'_N = \Delta H'_X + Y'_{Q/X} \tag{49}$$

This represents a simplified enthalpy balance for the formation of one C-mole of biomass. By dividing Eq. (49) by its left-hand side, one obtains:

$$1 = \frac{\Delta H'_X}{\Delta H'_S/Y'_{X/S} + x_3\,\Delta H'_N} + \frac{Y'_{Q/X}}{\Delta H'_S/Y'_{X/S} + x_3\,\Delta H'_N} \tag{50}$$

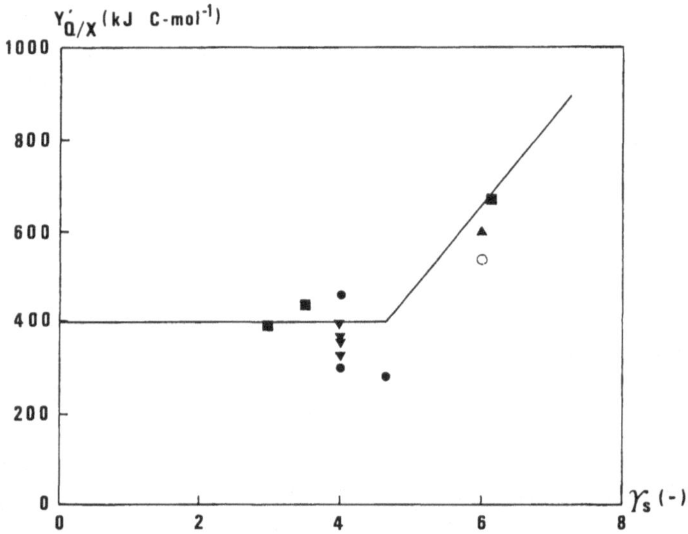

Fig. 16. Relationship between the heat yield $Y'_{Q/X}$ and the reductance degree of the substrate [57]

The first right-hand term of Eq. (50) represents the fraction of the chemical energy available in the substrate that is retained in the biomass. This fraction has been called the "enthalpy efficiency of growth" η_H by Roels [70]:

$$\eta_H = \frac{\Delta H'_X}{\Delta H'_S/Y'_{X/S} + x_3\,\Delta H'_N} \tag{51}$$

The second term shows the fraction of the energy dissipated as heat and can be measured by calorimetry. As suggested by von Stockar et al. [60], it is possible to deduce η_H directly from $Y'_{Q/X}$ by substituting Eq. (51) into Eq. (50):

$$\eta_H = \frac{\Delta H'_X}{Y'_{Q/X} + \Delta H'_X} \tag{52}$$

Values for enthalpic growth efficiency that were determined by calorimetry are plotted as a function of γ_s in Fig. 17. The line shows a theoretical prediction [60] obtained by substituting Eq. (47) and Eq. (19) into the definition (51) of η_H.

Both the experimental values and the theoretical prediction confirm that microorganisms are not able to retain in the biomass more than about 55–65% of the energy available in the rather oxidized substrates appearing on the left-hand side of the diagram. About 40% must be dissipated as heat in order to drive the growth process at a non-zero rate and to account for inefficiencies. Because of this limitation, and because of the low energy content of substrates with $\gamma_s < \gamma_s^c$, not much biomass can be formed, and $Y'_{X/S}$ stays below the maximum value of 0.6 (Eq. 47a). More reduced, i.e., more energy-rich substrates afford a higher yield. But as the degree of reduction of substrates exceeds γ_s^c in the right part of the diagram, an upper limit

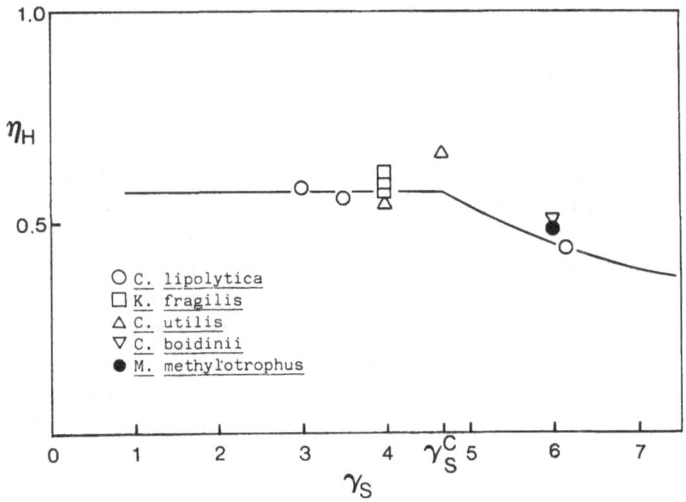

Fig. 17. Relationship between the enthalpic growth efficiency η_H and the reductance degree of the substrate γ_S determined calorimetrically. (Re-drawn with permission from [60])

of $Y'_{X/S}$ is reached as the organisms are not able to retain in the biomass more than about 60% of the carbon atoms available in the substrate. The excess energy in highly reduced substrate is thus released as heat and the enthalpic growth efficiency η_H decreases whereas $Y_{Q/X}$ increases.

4.5 Oxygen Uptake

It was demonstrated in Sect. 3.3 (Eqs. (28 and 29)) that the amount of heat ($Y'_{Q/O}$) generated by aerobic growth processes would be expected to amount to about 460 kJ per mole of oxygen consumed regardless of microbial strain and nature of substrate or product. In their pioneering study, Cooney et al. [24] grew a range of different organisms on both defined and complex media containing a carbon source of intermediate reduction ($\gamma_s = 4$). During eight growth experiments, they observed an average value of 518 ± 12 kJ (mol O_2)$^{-1}$. Working with a wider range of substrates with respect to their degrees of reduction, Volesky et al. [95] found 465 ± 84 kJ mol^{-1}, whereas Birou et al. [57] report 440 ± 33 kJ mol^{-1} (see Table 6).

In cultures with low oxygen uptake rates (OUR) and thus low aerobicity, the ratio of heat production to oxygen consumption is expected to increase towards infinity as shown by Eq. (43) in Table 4. No one has ever verified this equation experimentally by varying the degree of aerobicity systematically from $\Omega = 1$ to $\Omega = 0$, but Volesky et al. [95] reported values for $Y'_{Q/O}$ as high as 1710 kJ mol^{-1} for growth of *S. cerevisiae* when respiration was partially repressed by a glucose effect. At any rate, a complete verification of Eq. (43) would be demanding because the low OUR occurring at low values of Ω are difficult to measure precisely.

It can be concluded that in aerobic cultures, heat evolution rates and oxygen uptake rates bear the same information [74–77] and that an on-line measurement of both parameters will not contribute new information. This does, by no means, render heat flux measurements superfluous. On an industrial scale, they can be used as an independent check of oxygen uptake measurements and will be highly valuable as a basis for increased reliability of the overall process monitoring as well as for improving measurement accuracy owing to the availability of two independent sources of information. At pilot or bench scale, it is often not possible to measure oxygen uptake rates accurately due to low oxygen depletions resulting from high aeration rates. In this case, measuring heat dissipation rates and determining the OUR indirectly by means of Eq. (29) might well prove to be the most accurate alternative.

Finally, in aerobic processes involving facultative anaerobes, and which might become partially anaerobic, and in all processes involving non-respirative catabolic reactions such as often encountered in animal cell cultures heat evolution and oxygen consumption decouple and thus reflect different aspects of the metabolism. Measuring both of them simultaneously will then greatly improve the information on the state of the culture.

4.6 Product Formation

Primary and secondary product formation can be readily determined from thermograms. Thus acetate, a primary product of anaerobic growth, was detected during the aerobic growth of *E. coli* as a second peak in the heat signal [30,47] (Figs. 9a and b). This peak corresponds to the non-growth related oxidation of acetate formed during the initial growth period. By varying the initial glucose concentration supplied to the cultures, it was possible to show that purely respiratory metabolism, that is, the absence of product formation, only occurred when the initial glucose concentration was less than 0.75 g L^{-1} (Fig. 9c). Belaich [96] working with anaerobic glucose-limited batch cultures of *E. coli* showed that a second heat peak was observed unter certain pH conditions. Although the nature of the primary metabolic product (the oxidation of which gave rise to this peak) was not shown conclusively, mutants lacking the enzyme hydrogen-lyase did not exhibit this peak, thus indicating the production of formate or related product (Figs. 18a & b).

The energy balance described by Eqs. (25) and (26) contains a term which allows for the formation of bioproducts. If the value of $Y_{Q/X}$ is known, then Eq. (26) can be rearranged for the determination of the energy recovery (ER) in the same way as Eq. (50).

$$ER = \frac{Y'_{X/S}\,\Delta H'_X + \sum_i Y'_{P_i/S}\Delta H'_{P_i} + Y'_{X/S}Y'_{Q/X}}{\Delta H'_S + x_3 Y'_{X/S}\,\Delta H'_N} \tag{53}$$

If all of the products have been taken into account, then the energy recovery (ER) should have a value of unity. Values of ER significantly lower than 1 signify the formation of some undetected product [47]. If the nature of the product is known, the heat yield can serve together with $Y'_{X/S}$ to compute the amount of product formation by solving Eq. (26) for $Y'_{P/S}$ [4].

Relatively little attention has been focused on calorimetric investigations of secondary metabolite formation although this represents an important potential

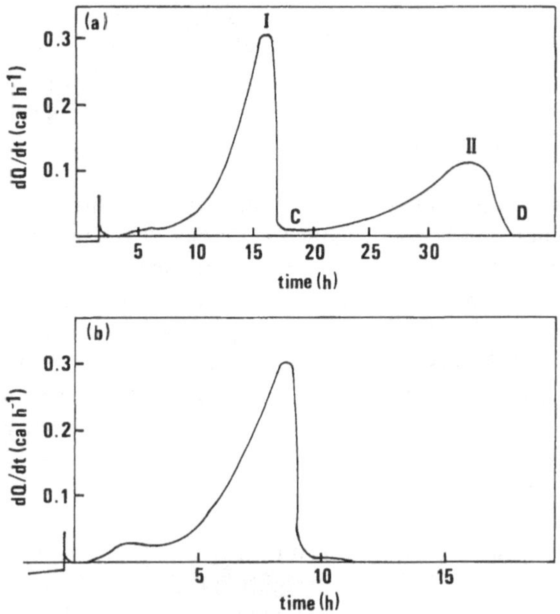

Fig. 18. Thermograms of *E. coli* k-12 322 (*a*) and *E. coli* 376 p4x (hydrogen-lyase mutant) (*b*) growing anaerobically in an anaerobic glucose-limited batch culture. (Re-drawn with permission from [96])

application due to the ability to detect and measure non-growth related events. Mou and Cooney [25] showed how "dynamic calorimetry" could be used for the monitoring and control of novobiocin and cellulose processes involving *Streptomyces niveus* and *Trichoderma viride*. In the former case growth and total heat evolved were found to be directly related until growth became limited by the nitrogen source. At this point the heat signal (q, $W\ L^{-1}$) fell to a value higher than the basal level and remained constant due to the production of novobiocin. Mou and Cooney [25] proposed that integration of this heat signal yields the maintenance requirement for the novobiocin process and that the amount of this product formed could be directly measured from the heat product yield coefficient $Y_{Q/P}$ (kJ heat evolved per g novobiocin formed). In batch experiments with both *S. niveus* and *T. viride*, the complex medium contained solids which made biomass and product determination difficult. Heat measurements are not affected by the presence of solids and only require regular calibration in order to correct the signal for changes in the rheological properties of the medium, which is readily possible in "macro" or bench-scale calorimeters.

Such experiments show the direct application of calorimetry in the identification of specific metabolic events, including the formation of non-growth related product formation and oxidation.

The most rigorous method discussed thus far that permits the use of heat dissipation measurements to detect product formation, consists in looking for and quantitatively analyzing enthalpy balance defects. A last case shall now be presented in which the growth processes can be modelled as demonstrated in Sect. 3.4, such that determination of one single heat yield (e.g. $Y'_{Q/X}$) is sufficient for quantitatively monitoring the extent of product formation in terms of a yield ($Y'_{P/S}$) as well as the complete stoichiometry of the process.

Examples of such cases include facultative anaerobes, such as yeasts grown under partially oxygen-limited conditions or under conditions of a "glucose effect". As it is conceivable on a large-scale industrial size to measure a heat yield on-line without having to resort to sophisticated sterilizable sensors, respective industrial processes such as the production of baker's yeast could potentially be controlled based on this measurement.

In order to verify the validity of the model described in Sect. 3.4, the yeast *K. fragilis* was grown as a continuous culture in a bench-scale calorimeter [59]. By limiting the oxygen supply to the culture, it was possible to obtain steady state growth with mixed metabolism between respiration and pure anaerobic growth. By varying the oxygen limitation, the mix could be shifted from pure respiration with no product formation but with a high biomass yield to pure anaerobic metabolism with full production of ethanol and low $Y_{Q/X}$. The biomass, product and heat yields were measured at the steady states representing these two extremes but also at many different steady states representing intermediate situations with different degrees of respiration and anaerobic metabolism.

This study confirmed the theoretical predictions set forth in Sect. 3.4 quite well. For instance, the heat generated per unit biomass formed ($Y'_{Q/X}$) varied in one chemostat culture from a mean value of 13.38 kJ g^{-1} (dry weight) (344 kJ C-mol^{-1}) for pure aerobic to 4.4 kJ g^{-1} (114 kJ C-mol)$^{-1}$ for pure anaerobic metabolism, whereas the theoretical values predicted by Eq. (42) in Table 4 were 13.6 kJ g^{-1} (350 kJ C-mol^{-1}) and 3.7 kJ g^{-1} (95.2 kJ C-mol^{-1}), respectively. As the aerobicity Ω of the culture was decreased from 1 to 0 by limiting the oxygen supply, $Y'_{Q/X}$ decreased continuously in a hyperbolic manner roughly in accord with Eq. (42).

Figure 19 illustrates how monitoring the heat generation might be used to calculate on-line the bioproduct formation and the whole state of the culture. Let us assume that $Y'_{Q/S}$, the ratio of heat release to substrate consumption, could be

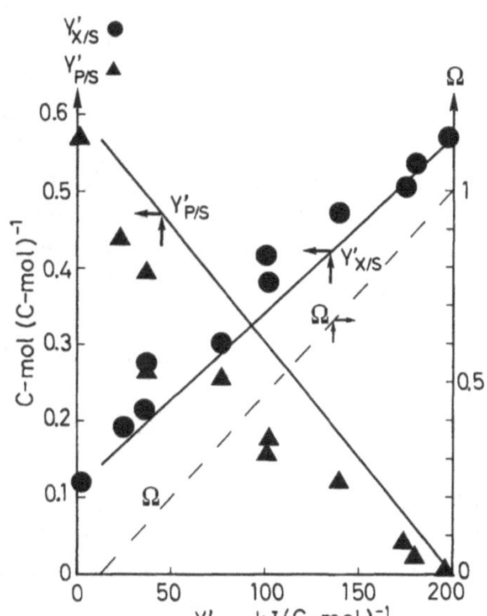

Fig. 19. Determination of yield coefficients and 'degree of aerobicity' Ω, of a culture from calorimetric measurement of heat yields (Eq. 41, 54 and 55) (●) C-molar biomass yield $Y'_{Q/S}$ and (▲) C-molar product yield $Y'_{X/S}$ as a function of the heat yield $Y'_{Q/S}$

measured continuously. The measured value of $Y'_{Q/S}$ could be used to determine Ω according to Eq. (41) in Table 4, which appears as the *straight broken line* in Fig. 19. As the whole state of the culture depends on its aerobicity Ω, all other yields can be evaluated from there. $Y'_{P/S}$ and $Y'_{X/S}$ depend on Ω as follows [59]:

$$Y'_{P/S} = \frac{\gamma_s - \gamma_x (Y'_{X/S})_F}{\gamma_p} (1 - \Omega) \tag{54}$$

$$Y'_{X/S} = (Y'_{X/S})_R \, \Omega + (Y'_{X/S})_F (1 - \Omega) \tag{55}$$

Using values for the parameter $(Y'_{X/S})_F$ and $(Y'_{X/S})_R$ given in Ref. [59], the product and biomass yields were computed by Eqs. (54) and (55) and plotted as a function of $Y'_{Q/S}$ (*straight solid lines*).

As both $Y'_{P/S}$ and $Y'_{X/S}$ were measured and the heat yield $Y'_{Q/S}$ can be evaluated from Ref. [59], it is possible to compare indirect determinations of $Y'_{P/S}$ and $Y'_{X/S}$ (*solid lines* on Fig. 19) based on $Y'_{Q/S}$ with actually measured values (*triangles* and *dots*). The experimental biomass yields (*dots*) agree reasonably well with the calculations, whereas the experimental ethanol yields appear systematically low. This was very probably due to partial stripping of the ethanol from the chemostat culture and it is conceivable that the $Y'_{P/S}$ determinations based on heat yields actually represent better estimates than those derived from measured ethanol concentrations.

Similar determinations can be done using $Y'_{Q/C}$, the ratio of dissipated heat to CO_2 generation, as heat yield. This is especially appropriate as CO_2 evolution rates are routinely measured on-line. Results obtained by using $Y'_{Q/C}$ were very similar to Fig. 19a [97].

Figure 19 illustrates a case in which the whole state of a microbial culture can be deduced from the measurement of one single heat yield.

5 Conclusions

Section 4 of this report shows that microbial heat generation depends in very specific ways on growth, biomass yield, maintenance metabolism, nature of substrate and energetic efficiency of growth, oxygen uptake, and product formation. Conversely, it is possible to resort to heat measurements to gain information on any of these factors.

Thus, a considerable amount of information regarding the growth of microbial cultures can be determined from thermograms. The shape of thermograms obtained with microbial cultures grown under defined batch culture conditions on specific carbon and energy sources has been proposed as a taxonomic method for the indentification and "finger-printing" of microorganisms. Thermograms lend themselves in an ideal way for monitoring general microbial activity and for observing and pinpointing all kinds of metabolic events such as product formation with subsequent oxidation, medium limitations, inhibition, and diauxies. As shown in Sect. 4, these features could be exploited in a bench-scale calorimeter in order to optimize growth media, culture conditions, and product formation. The fast response of the heat flux calorimeter and its direct correlation with biomass formation also suggest the possibility of convenient and rapid determinations of microbial growth kinetics. In strongly aerobic growth process, a more or less general correlation between heat generation and oxygen consumption exists which is used to crosscheck or, under certain conditions, to substitute the oxygen uptake by heat flux measurements. The advantages of the latter over the former are improved accuracy at the bench scale and considerably faster response times. As a conclusion, "macro"-calorimetry is expected to become a valuable tool for process development work.

On a more fundamental level, calorimetry may be used to study energetic and enthalpic growth efficiency as well as maintenance energy dissipation, thereby offering new insight into microbial growth energetics. Some of this research is based on calorimetric chemostat experiments at the bench scale, but it is evident that the potential of this technique has only just begun to be explored and might open up a host of new research opportunities. Chemostat experiments in bench-scale calorimeters could thus be used to study spontaneous synchronization, and oscillating cultures and the like whilst transient responses to various types of step changes could provide a novel tool for investigating inhibition, induction and repression kinetics. All of these applications point to the role calorimetry has yet to play in more basic research.

It has also been demonstrated that heat dissipation measurements on industrial scale hold quite some potential as a basis for process control. In contrast to other process variables, the heat generated by the culture can be measured on-line without sophisticated probes.

6 Symbols

A	heat transfer area (Eq. (4)) (m^2)
A_i	ash fraction of compound i
A_j	ash fraction of compound j
C	carbon
D	dilution rate (h^{-1})
ER	energy recovery (Eq. (53))
H	hydrogen (1)
$\Delta H_i'$	molar or C-molar heat of combustion of compound i, kJ mol^{-1} or kJ C-mol^{-1}
ΔH_i^*	heat of combustion of compound i assuming nitrogen is released as NH_3 (Eq. (23)), kJ C-mol^{-1}
k	correction factor for adiabatic calorimetry (Eq. (1)) ideally equal to heat capacity of vessel and contents (J K^{-1}) (Eq. (1))
K_p	total heat capacity of system (J K^{-1}) (Eq. (2))
m_Q	heat evolved due to maintenance (W g^{-1}) (Eq. (44))
M_i'	formula mass of one C-mol of compound i
M_j'	formula mass of one C-mol of compound j
N	nitrogen (3)
O	oxygen (2)
OUR	oxygen uptake rate (mol L^{-1} h^{-1})
P_1, P_2, P_3	moles of element H, O, and N per C-mol product (mol C-mol^{-1})

q	heat evolution rate ($W\ L^{-1}$)
q_A	heat generated through agitation (W)
q_C	cooling power (Eq. (3)) (W)
q_F	heat transferred to cooling system (Eq. (4)) (W)
q_G	heat lost to gas-stream (W)
q_H	heat released by electrical heater (Eq. (3)) (W)
q_L	heat losses from the system (W)
q_R	heat generated by the reaction (W)
Q	heat liberated (Eq. (1)) (kJ)
Q_0	Heat evolved per equivalent of electrons transferred to oxygen (given by Eq. (19)) (kJ $C\text{-}mol^{-1}$ eq^{-1} electrons)
r'_i	conversion rate of compound i ($C\text{-}mol\ s^{-1}\ L^{-1}$)
r'_j	conversion rate of compound j ($C\text{-}mol\ s^{-1}\ L^{-1}$)
S_1, S_2, S_3	moles of element H, O, and N per C-mol substrate S ($mol\ C\text{-}mol^{-1}$)
T_J	temperature of jacket oil (Eq. (4)) (K)
T_R	temperature of reactor contents (Eq. (4)) (K)
U	global heat transfer coefficient (Eq. (4)) ($Wm^{-2}\ K^{-1}$)
V	potential difference (volts)
W	water
X_1, X_2, X_3	moles of element H, O or N per C-mol biomass X ($mol\ C\text{-}mol^{-1}$)
$Y_{i/j}^{max}$	maximum yield coefficient of i with respect to j for $m_Q = O$ (Eq. (44)), ($g\ g^{-1}$)
$Y_{i/j}^{min}$	minimum yield coefficient of i with respect to j (Eq. (45)) ($g\ g^{-1}$)
Y_{kJ}	specific heat yield coefficient ($kJ\ g^{-1}$ cell dry weight)
$Y_{Q/i}$	heat yield coefficient with respect to compound i ($kJ\ g^{-1}$)
$Y_{Q/X}$	heat yield coefficient with respect to biomass ($kJ\ g^{-1}$)
$Y'_{C/S}$	CO_2 yield coefficient ($mol\ C\text{-}mol^{-1}$)
$Y'_{N/S}$	Nitrogen yield coefficient ($mol\ C\text{-}mol^{-1}$)
$Y'_{O/S}$	Oxygen yield coefficient ($mol\ C\text{-}mol^{-1}$)
$Y'_{P/S}$	Product yield coefficient ($mol\ C\text{-}mol^{-1}$)
$Y'_{Q/C}$	heat yield coefficient with respect to CO_2 ($kJ\ C\text{-}mol^{-1}$)
$Y'_{Q/O}$	heat yield coefficient with respect to oxygen ($kJ\ mol^{-1}$)
$Y'_{Q/S}$	heat yield coefficient with respect to substrate ($kJ\ C\text{-}mol^{-1}$)
$Y'_{X/S}$	Biomass yield coefficient ($mol\ C\text{-}mol^{-1}$)
$Y'_{W/S}$	H_2O yield coefficient ($mol\ C\text{-}mol^{-1}$)
$(Y'_{i/j})_F$	yield coefficient of component i with respect to j for purely anaerobic growth (Eq. (32)) ($mol\ C\text{-}mol^{-1}$)
$(Y'_{i/j})_R$	yield coefficient of component i with respect to j for pure respiratory growth (Eq. (31)) ($mol\ C\text{-}mol^{-1}$)

Greek Symbols:

$\delta_s, \delta_x, \delta_p$	degree of hydrogenation of substrate, biomass and product, respectively
γ_s^c	constant = 4.67 (Eq. (47))
γ_i^0	reductance degree of compound i with respect to N_2
$\gamma_s, \gamma_x, \gamma_p$	reductance degree of substrate, biomass and product respectively defined with respect to NH_3

η_H enthalpy efficiency of growth (Eq. (51))

μ specific growth rate (h^{-1})

Ω 'degree of aerobicity' of culture (Eq. (33)) i.e., the relative fraction of aerobic to anaerotic metabolism. $\Omega = 1$ for purely aerobic growth, $\Omega = 0$ for purely anaerobic growth

7 References

1. Lavoisier AL, de Laplace PS (1784) Histoire de l'Académie des Sciences, Année 1780 p 355
2. Jéquier E (1983) Nestlé Research News *1982/83*: 27
3. Birou B, von Stockar U (1989) Enz. Microb. Technol. *11*: 12
4. Marison IW, von Stockar U (1987) Enz. Microb. Technol. *9*: 33
5. Dubrunfaut M (1856) C.R. Acad. Sci. Paris *42*: 945
6. Rubner M (1903) Hyg. Rundsch. *13*: 857
7. Rubner M (1903) Arch. Hyg. *48*: 260
8. Rubner M (1904). Arch. Hyg. *66*: 81
9. Bouffard A (1895) C.R. Acad. Sci. Paris *121*: 357
10. Tian A (1923) Bull. Soc. Chim. *33*: 427
11. Calvet E (1948) C.R. Acad. Sci. Paris *226*: 1702
12. Wadsö I (1968) Acta Chem. Scand. *22*: 927
13. Fujita T, Nunomura K, Kagami I, Nishikawa Y (1976) J. Gen Microbiol. *22*: 43
14. Lamprecht I, Schaarschmidt B (1973) Bull. Soc. Chim. Fr. *4*: 1200
15. Eriksson R, Holme T (1973) Biotechnol. Bioeng. Symp. *4*: 581
16. Forrest WW, Walker DJ, Hopgood MF (1961) J. Bacteriol. *82*: 685
17. Belaich JP, Senez JC, Murgier M (1968) J. Bacteriol. *95*: 1750
18. Beezer AE (1976) In: Lamprecht I, Schaarschmidt B (eds) Application of calorimetry in life sciences. de Gruyter, Berlin. p. 109
19. Wadsö I (1980) In: Beezer AE (ed.) Biological microcalorimetry, Academic Press, London p. 247
20. Ishikawa Y, Shoda M, Maruyama H (1981) Biotechnol. Bioeng. *23*: 2629
21. Ishikawa Y, Shoda M (1983) Biotechnol. Bioeng. *25*: 1817
22. Demoun Z, Boussand R, Cotten D, Belaich J-P (1985) Biotechnol. Bioeng. *27*: 996
23. Demoun Z, Belaich J-P (1985) Biotechnol. Bioeng. *27*: 1005
24. Cooney CL, Wang DIC, Mateles RI (1968) Biotechnol. Bioeng. *11*: 269
25. Mou DG, Cooney CL (1976) Biotechnol. Bioeng. *18*: 1371
26. Wang H, Wang DIC, Cooney CL (1978) Eur. J. Appl. Microbiol. *5*: 207
27. Volesky B, Luong JHT, Thambimuthu KV (1978) Can. J. Chem. Eng. *56*: 526
28. Luong JHT, Volesky B (1980) Can. J. Chem. Eng. *58*: 497
29. Luong JHT, Yerushalmi L, Volesky B (1983) Enz. Microb. Technol. *5*: 291
30. Marison IW, von Stockar U (1986) Biotechnol. Bioeng. *28*: 1780
31. Marison IW, von Stockar U (1985) Thermochim. Acta *85*: 493
32. Karlsen LG, Villadsen J (1987) Chem. Eng. Sci. *42*: 1153
33. Hemminger W, Höhne G (1984) In: Calorimetry. Fundamentals and practice, Verlag Chemie, Weinheim, FRG. p 131
34. Wadsö I (1987) In: James AM (ed) Thermal and energetic studies of cellular biological systems. Wright, Bristol, UK. p 34
35. Spink CH (1980) CRC Crit. Rev. Anal. Chem. *9* (1): 1
36. Martin CJ, Marini MA (1979) CRC Crit. Rev. Anal. Chem. *8* (3): 221
37. Sturtevant JM (1971) In: Weissberger A, Rossiter BW (eds) Physical methods of chemistry vol I part 5. John Wiley, New York. p 347
38. Wadsö I (1975) In: Pain R, Smith B (eds) New Techniques in Biophysics and Cell Biology vol 2. R. John Wiley, New York. p 85
39. Spink C, Wadsö I (1976) In: Glick D (ed) Methods of biochemical analysis vol 23. John Wiley, New York, p 1

40. Kubaschewski O, Hultgren R (1962) In: Skinner HA (ed) Experimental thermochemistry vol 2. Interscience, New York
41. Calvet E (1953) C.R. Acad. Sci. Paris *236*: 377
42. Calvet E, Prat H (1956) Microcalorimétrie, applications physiochimiques et biologiques. Masson, Paris
43. Calvet E, Prat H (1963) Recent progress in microcalorimetry. Pergamon, London, U.K.
44. Schaarschmidt B, Reichert U (1981) Exp. Cell. Res. *131*: 480
45. Martin CJ, Marini MA (1979) CRC Crit. Rev. Anal. Chem. *8* (3): 221
46. Jarret IG, Clark DG, Filsell OH, Harvey JW, Clark MG (1979) Biochem. J. *180*: 631
47. Ishikawa Y, Nonoyama Y, Shoda M (1981) Biotechnol. Bioeng. *23*: 2825
48. Demoun Z, Belaich J-P (1979) J. Bacteriol. *140*: 377
49. Monk P, Wadsö I (1968) Acta Chem. Scand. *22*: 1842
50. Spink C, Wadsö I (1976) In: Glick D (ed) Methods of biochemical analysis. John Wiley, New York. p 1
51. Poore VM, Beezer AE (1983) Thermochim. Acta *63*: 133
52. Bayer K, Fuehrer F (1982) Process Biochem. *17*: 42
53. Luong JHT, Volesky B (1982) Eur. J. Appl. Microbiol. Biotechnol. *16*: 28
54. Regenass W (1977) Thermochim. Acta *20*: 65
55. Regenass W (1983) Chimia *37*: 430
56. Giger G, Aichert A, Regenass W (1982) Swiss Chem. *4* (3 a): 33
57. Birou B, Marison IW, von Stockar U (1987) Biotechnol. Bioeng. *30*: 650
58. Birou B (1986) PhD Thesis, EPFL, Switzerland No. 612
59. von Stockar U, Birou B (1987) Biotechnol. Bioeng. (accepted for publication)
60. von Stockar U, Marison IW, Birou B (1987) In: Moody GW, Baker PB (eds) Bioreactors and biotransformations. Elsevier, London, p 87
61. Sedlaczek L (1964) Acta Microbiol. Polon. *13*: 101
62. Prochazka GJ, Payne WJ, Mayberry WR (1970) J. Bacteriol. *104*: 646
63. Prochazka GJ, Payne WJ, Mayberry WR (1973) Biotechnol. Bioeng. *15*: 1007
64. Belaich JP (1980) In: Beezer AE (ed.) Biological Microcalorimetry. Academic, London. p 1
65. Ho KP, Payne WJ (1979) Biotechnol. Bioeng. *21*: 787
66. Cordier JL, Butsch BM, Birou B, von Stockar U (1987) Appl. Microbiol. Biotechnol. *25*: 305
67. Thornton WM (1917) Philos. Mag. *33*: 196
68. Kharasch MS, Sher B (1925) J. Phys. Chem. *29*: 625
69. Giese AC (1968) Cell Physiology, 3rd Edition. W. B. Saunders, Philadelphia, p 412
70. Roels JA (1983) Energetics and kinetics in biotechnology. Elsevier Biomedical, Amsterdam p 330
71. Minkevich IG, Eroshin VK (1973) Biotechnol. Bioeng. Symp. *4*: 21
72. Erickson LE, Minkevich IG, Eroshin VK (1978) Biotechnol. Bioeng. *20*: 1595
73. Erickson LE, Selga SE, Viesturs UE (1978) Biotechnol. Bioeng. *20*: 1623
74. Stephanopoulos G, San KY (1984) Biotechnol. Bioeng. *26*: 1176
75. San KY, Stephanopoulos G (1984) Biotechnol. Bioeng. *26*: 1189
76. Grosz R, Stephanopoulos G (1984) Biotechnol. Bioeng. *26*: 1198
77. Stephanopoulos G, San KY (1984) Biotechnol. Bioeng. *26*: 1209
78. Russell WJ, Zettler JR, Blanchard GC, Boling AE (1975) In: Heden C-G, Illeni T (eds) New approaches to the identification of microorganisms, John Wiley, London, p 101
79. Ljungholm K, Wadsö I, Mardh P-A (1976) J. Gen. Microbiol. *96*: 283
80. Monk P, Wadsö I (1975) J. Appl. Bacteriol. *38*: 71
81. Perry BF, Beezer AE, Miles RJ (1983) J. Appl. Bacteriol. *54*: 183
82. Schaarschmidt B, Lamprecht I (1976) Experienta *32*: 1230
83. Beezer AE, Bettelheim KA, O'Farrell SM, Al-Salihi S, Shaw EJ (1977) In: Johnson HH, Newson SWB (eds) 2nd Int. Symp. Rapid Methods and Automation in Microbiology. Learned Information, Oxford
84. Lamprecht I, Meggers C (1969) Z. Naturforsch. (C) *24b*: 1205
85. Bowden CPP, James AM (1985) Microbios *44*: 75
86. Forrest WW, Walker DJ, Hopgood MF (1961) J. Bacteriol. *82*: 648
87. Luong JHT, Volesky B (1982) Can. J. Chem. Eng. *60*: 163
88. Luong JHT, Volesky B (1983) Adv. Biochem. Eng. *28*: 1

89. Nagai S, Aiba S (1972) J. Gen. Microbiol. *73*: 531
90. Poole AK, Haddock BA (1975) FEBS Letts. *58*: 249
91. Brettel R, Lamprecht I, Schaarschmidt B (1981) Eur. J. Appl. Microbiol. Biotech. *11*: 201
92. Linton JD, Stephanson RJ (1978) FEMS Microbiol. Letts. *3*: 95
93. Roels JA (1980) Biotechnol. Bioeng. *22*: 2457
94. Heijnen JJ, Roels JA (1981) Biotechnol. Bioeng. *23*: 739
95. Volesky B, Yerushalmi L, Luong JHT (1982) J. Chem. Technol. Biotechnol. *32*: 650
96. Belaich A, Belaich J-P (1976) J. Bacteriol. *125*: 14
97. von Stockar U, Marison IW: The use of on-line heat measurements in bioreactor control. (To be published)
98. Birou B, Marison IW, von Stockar U: Proc. 4th Europ. Congr. Biotech. (1987) vol 3, p 105, Elsevier, Amsterdam

Effect of Total and Partial Pressure (Oxygen and Carbon Dioxide) on Aerobic Microbial Processes

U. Onken and E. Liefke
Universität Dortmund, Fachbereich Chemietechnik, Lehrstuhl Technische Chemie B.
Postfach 500 500, D-4600 Dortmund 50

1 Introduction . . . 138
2 Effects of Total Pressure . . . 139
2.1 Higher Pressures (up to ca 1000 bar) . . . 139
2.2 Moderate Pressures . . . 140
3 Carbon Dioxide Partial Pressure . . . 141
3.1 Dissolved Carbon Dioxide and Ionic Equilibria . . . 141
3.2 Effect of Carbon Dioxide on Microbial Growth . . . 143
4 Oxygen Partial Pressure . . . 145
4.1 Dissolved Oxygen Concentration — Intrinsic Parameter . . . 145
4.2 Analysis of Effects of Total Pressure and Oxygen Partial Pressure . . . 147
4.2.1 Experimental Techniques . . . 147
4.2.2 Procedure and Analysis of Experiments . . . 149
4.3 Effect of Oxygen on Microbial Cultures . . . 151
4.3.1 Growth . . . 151
4.3.2 Formation of Products . . . 153
4.3.3 Possible Causes of Toxic Oxygen Effects on Growth . . . 156
5 Timed Variations of Oxygen Partial Pressure . . . 159
5.1 Experimental Techniques . . . 159
5.2 Effects of Single Shifts . . . 161
5.3 Effects of Cyclic Variations . . . 163
6 Conclusions . . . 165
7 Acknowledgments . . . 165
8 Symbols . . . 166
9 References . . . 166

In industrial bioreactors, levels and gradients of total and partial pressures are considerably higher than on the laboratory scale. In the relevant range (in general up to 2 or 3 bar, maximum approx. 10 bar), effects of total pressure on aerobic cultures are neglegibly small. CO_2 partial pressures of more than approx. 100 mbar may have inhibitory effects on aerobic cultures. Growth of aerobic cultures can be enhanced by O_2 partial pressures higher than 210 mbar (corresponding to air at 1 bar), if oxygen transfer is limited. In many cases, however, increased O_2 partial pressure (higher than approx. 1 bar) is toxic to aerobic cultures and inhibits microbial growth and product formation. Stepwise and cyclic variations of O_2 partial pressure may have positive or negative effects, depending on strain of microorganism, culturing conditions, and range of dissolved oxygen concentration.

Knowledge of these effects is required in process development and bioreactor scale-up.

1 Introduction

In microbial processes, the metabolic capabilities of microorganisms are used for the conversion of reactants into products, such as the manufacture of metabolites, the transformation or the degradation of chemicals, or even the production of protein in the form of the microorganisms themselves. Since in all these processes, the microorganisms play the main part, their growth is of predominant importance for process operation. Kinetics of microbial growth is influenced by a great multitude of parameters. The majority of these are environmental parameters characteristic of the chemical and physical environment of cells, e.g. concentrations of substrate, ions, dissolved gases, temperature, pressure, shear stress. Apart from these variables, the physiological state of the culture, including the distribution of cell ages, affects microbial kinetics; here in particular the history of the individual cell may be significant.

Investigations of microbial processes are generally carried out under controlled, if possible under constant environmental conditions. In batch cultures, of course, several parameters, e.g. concentrations of substrates, metabolites, and dissolved gases, will change with time; however spatially, that is at any position of the system, they attain practically constant values. On laboratory scale with fluid systems, that is in laboratory bioreactors, this is usually achieved easily by efficient mixing. However, with increase in size uniform distribution of concentrations and temperature becomes more difficult, especially in industrial reactors in continuous operation with a single feeding point. Still, by distributing a larger number of feeding points and good mixing, spatial concentration gradients can be smoothed.

With dissolving gases fed into the reactor, the situation is quite different, because in large bioreactors of great height hydrostatic pressure is not constant throughout the reactor, but a function of the local position, i.e. of liquid height. This is of course also true of total pressure as the sum of hydrostatic pressure and pressure at the reactor top. As a consequence, equilibrium solubilities of gaseous compounds will also be a function of local position in the reactor. Thus, in an industrial bioreactor containing a liquid medium up to a level of 10 m and with a pressure at the top of 1.5 bar abs., total pressure at the bottom will amount to 2.5 bar abs.; this means, that gas solubility, e.g. for oxygen or carbon dioxide at the bottom, will be nearly 70% higher than at the top of the bioreactor. This difference in gas solubility is clearly much larger in very high bioreactors with liquid heights of 60 m in the ICI Pressure Cycle Fermenter for sterile bioprocesses, developed for Single Cell Protein from methanol [1,2,3], up to 30 m in the BIOHOCH Reactor HOECHST [4] and the BAYER Tower Biology Reactor [5], and more than 100 m for the ICI Deep-Shaft Reactor [6]; these latter three types of bioreactors are used in aerobic effluent treatment.

Summing up, it can be stated, that in industrial bioreactors pressure is a function of the local position with average values definitely above 1 bar abs., whereas laboratory investigations of microbial processes are usually performed at atmospheric pressure. Total pressure directly affects partial pressures. Since these are especially important in the kinetics of aerobic processes, effects of both total and partial pressures will be presented and discussed in this survey.

2 Effects of Total Pressure

2.1 Higher Pressures (up to approx. 1000 bar)

Experimental investigations of the effects of total pressure on microorganisms cover a wide range of pressures. Characteristic results from these investigations are given in Fig. 1. They have been carried out by exerting pressure on specimen tubes without gassing the culture, e.g. by putting the specimen tubes into an autoclave which was subsequently pressurized. ZoBell and his co-workers were the first to perform such experiments [7,8,14].

It can be seen from Fig. 1, that many microorganisms can exist at rather high pressures. Though in some cases positive effects of hydrostatic pressure on microbial growth have been observed up to about 300 bar, there is a general tendency of growth inhibition by pressures much higher than the usual environmental pressure of the microorganisms.

The graph in Fig. 1, of course, shows a rough picture only. In a more detailed discussion of the inhibition of microorganisms by increased pressures, other environmental parameters, such as temperature, medium and duration, have to be considered. Thus experiments with a strain of *Escherichia coli* have shown, that inactivation by high hydrostatic pressure is apparently a function of time with first order in cell number [15]. Blocking of protein synthesis by pressure should be the main factor

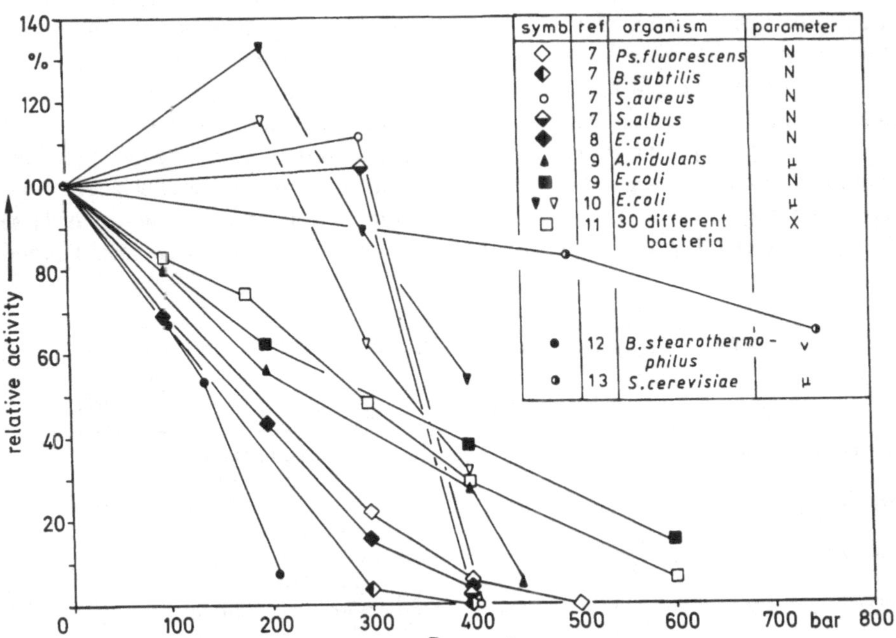

Fig. 1. Effect of total pressure *P* on microbial growth. Microbial activity is given by data for growth parameters related to values at 1 bar (relative activity)

responsible for stopping cell growth. More information on these and related aspects can be found in monographs and reviews [16,17,18].

Several species of microorganisms usually living at ambient pressure, i.e. 1 bar, show enough resistance to hydrostatic pressure to exist and grow well even at more than 400 to 500 bar. In contrast to these so-called barotolerant species, those microorganisms which have been isolated from the depths of the sea down to more than 10,000 m need high hydrostatic pressures of up to about 1000 bar as the optimum condition for growth. These barophilic microbes have apparently developed in ecological niches under extreme conditions. As the maximum pressure, at which such a barophilic microorganism has been found, a value of 1400 bar has been reported [10].

2.2 Moderate Pressures

Of course, the effects of high pressures of several hundred bar are not of relevance to aerobic processes, where in industrial bioreactors pressures of the order of up to 10 bar can only be reached locally. For this lower range of pressures a few investigations have been published regarding possible effects of pressure on microbial growth [6,19–22]. Results from these papers (see Table 1) are not conclusive. This is also true of observations made in industrial bioreactors. In the PROST reactor (PROST = Pressure Recycling Oxidation Sewage Treatment) the microorganisms are apparently not damaged in spite of cyclic pressure changes between 1 and 12 bar [23]. For the ICI Deep-Shaft reactor for aerobic waste water treatment respiration is reported to be considerably enhanced due to the cyclic pressure changes in the circulating liquid [24].

It is evident, that the reported effects of moderate pressures on aerobic microbial cultures cannot be easily explained, and that they are not simply a direct result of pressure (see footnote in Sect. 4.2.2). Obviously for a more detailed analysis of the pressure effects, the partial pressures of oxygen and of carbon dioxide must be considered. From these, oxygen partial pressure should be the candidate deserving more interest, because in aerobic processes oxygen is an indispensible nutrient. Besides partial pressures, another aspect has to be taken into account. In industrial bioreactors of large dimensions, especially great heights, microorganisms and surrounding medium under-

Table 1. Influence of moderate pressure on aerobic cultures

Organism	Pressure (bar)	Effect	Refs.
Pseudomonas aeruginosa	1–6	Decrease in cellular yield; decrease of μ_{max}	[19]
Candida utilis	1–8	Decrease in cellular yield; no effect on μ_{max}	[20]
Pseudomonas fluorescens	1–8	Growth inhibition (no growth at 8 bar)	[20,21]
Staphylococcus aureus	1–30	Growth inhibition	[22]
Escherichia coli	1–60 (pulsating)	Growth inhibition	[22]
Mixed culture (activated sludge)	1–10 (pulsating nitrogen)	No damage to cells	[6]

go more or less periodic changes of total pressure owing to the mixing in the reactor. These pressure changes will affect the concentrations of dissolved gases. Again, in aerobic cultures this will mainly concern oxygen. For discussing and elucidating effects of both constant and changing oxygen partial pressure a knowledge on effects of carbon dioxide partial pressure will he helpful and necessary. We will therefore treat CO_2 first, and subsequently oxygen.

3 Carbon Dioxide Partial Pressure

3.1 Dissolved Carbon Dioxide and Ionic Equilibria

Carbon dioxide is an important component in the metabolism of aerobic microorganism not only as a product, but also as an educt. The product CO_2 formed by decarboxylation reactions, e.g. in the respiration of organic compounds, will dissolve in the medium, from where it will be stripped by the flow of air supplied to the aerobic culture. On the other hand, CO_2 is needed for the production of intermediates necessary in metabolic sequences, such as oxalacetate which can be formed by carboxylation of pyruvate in five different ways. Thus part of the CO_2 from respiration may be consumed, but the overall balance of aerobic microbial growth on organics will yield a surplus of CO_2. This will lead to some over-saturation of the medium with CO_2, depending on the amount of CO_2 stripped by aeration.

In equilibrium with air containing 0,03 vol-% CO_2 the concentration of CO_2 in pure water is about 10^{-5} kmol m^{-3} (= 0,4 ppm) at 1 bar, 25 °C. In culture media the the solubility of CO_2 is generally somewhat lower because of their contents of dissolved organic substrates [25,26].

With CO_2 dissolved in aqueous media another fact has to be taken into account; dissolved CO_2 reacts with water yielding the ions hydrogen carbonate (HCO_3^-) and carbonate (CO_3^{2-}). They are the anions of carbonic acid (H_2CO_3); but since its concentration in equilibrium with dissolved CO_2 is very low (approx. 2×10^{-3} of CO_2 concentration), it may be neglected when the total quantity of dissolved CO_2 is of interest. This is not true for the concentrations of HCO_3^- and CO_3^{2-} formed by dissociation:

$$H_2CO_3 \rightleftharpoons H^+ + HCO_3^- \tag{1}$$

$$HCO_3^- \rightleftharpoons H^+ + CO_3^{2-} \tag{2}$$

For practical reasons the hydration of CO_2 according to

$$CO_2 + H_2O \underset{k_2}{\overset{k_1}{\rightleftharpoons}} H_2CO_3 \tag{3}$$

is usually included in the equilibrium constant of Eq. (1) which then reads:

$$K_1 = \frac{[H^+][HCO_3^-]}{[CO_2]} \quad (\approx 5 \times 10^{-7}\ \text{kmol m}^{-3} \text{ at } 25\ ^\circ\text{C}) \tag{4}$$

The equilibrium constant of Eq. (2) is:

$$K_2 = \frac{[H^+][CO_3^{2-}]}{[HCO_3^-]} \quad (\approx 5{,}6 \times 10^{-11}\ \text{kmol m}^{-3} \text{ at } 25\ ^\circ\text{C}) \tag{5}$$

The values for K_1 and K_2 given here are valid for pure water as the solvent; for culture media containing electrolytes and organic substrates they have to be corrected with

Table 2. Equilibrium concentrations of CO_2 dissociation at different CO_2 partial pressures in a nutrient medium, according to Eqs. (1—3); T = 30 °C, pH = 7.0, I = 0.131 kmol m^{-3}

p_{CO_2} [mbar]	$[CO_2]$ [kmol m^{-3}]	$[HCO_3^-]$ [kmol m^{-3}]	$[CO_3^{2-}]$ [kmol m^{-3}]
0.35	$1{,}02 \cdot 10^{-5}$	$1{,}09 \cdot 10^{-4}$	$2{,}90 \cdot 10^{-7}$
1.0	$2{,}90 \cdot 10^{-5}$	$3{,}11 \cdot 10^{-4}$	$8{,}30 \cdot 10^{-7}$
10.0	$2{,}90 \cdot 10^{-4}$	$3{,}13 \cdot 10^{-4}$	$8{,}52 \cdot 10^{-6}$
100.0	$2{,}90 \cdot 10^{-3}$	$3{,}44 \cdot 10^{-2}$	$1{,}13 \cdot 10^{-4}$
300.0	$8{,}71 \cdot 10^{-3}$	$1{,}32 \cdot 10^{-1}$	$7{,}13 \cdot 10^{-4}$

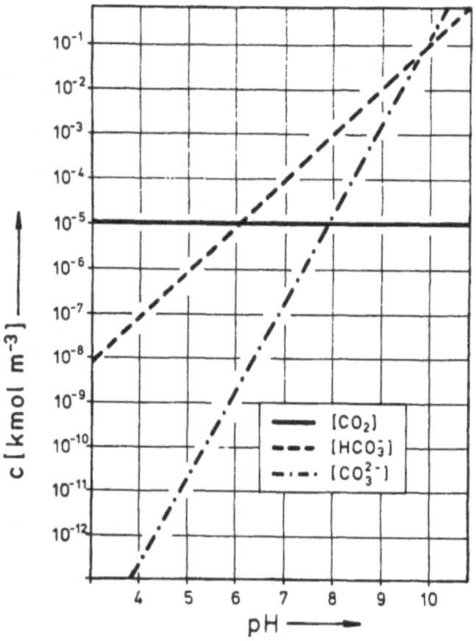

Fig. 2. Influence of pH on equilibrium between CO_2, HCO_3^- and CO_3^{2-} in synthetic nutrient medium (I = 0,131 kmol m^{-3}) at T = 30 °C and P_{CO_2} = 0,35 mbar, calculated according to Freier [27]

the aid of activity coefficients. Equilibrium concentrations for dissolved CO_2, HCO_3^- and CO_3^{2-} in a typical culture medium at 30 °C and pH = 7, which have been obtained with an activity coefficient function proposed by Freier [27], are given in Table 2. Under these conditions, evidently the major portion of CO_2 is present as HCO_3^- with nearly five times higher concentrations than CO_2, whereas the concentration of CO_3^{2-} is negligibly low. However, an inspection of Eqs. (4) and (5) shows that the equilibrium of CO_2 with HCO_3^- and CO_3^{2-} is strongly dependent on pH. This comes out impressively from Fig. 2. It shows that below pH 4 the ionic equilibria of CO_2 can be neglected. From pH 7 to 9 HCO_3^- is predominant; above pH 9 CO_3^{2-} must be included, becoming the controlling species at pH > 11.

Another point has to be considered, i.e. reaction velocity. Whereas the ionic equilibrium reactions, Eqs. (1) and (2), are very rapid, hydration of CO_2 (Eq. 3) is rather slow with $k_1 \approx 0.03\ s^{-1}$ and $k_2 \approx 20\ s^{-1}$ at 25 °C. This means that in stripping CO_2, the velocity of the dehydration step is fast enough to be neglected in comparison to the step of physical desorption from liquid to gas phase, whereas in dissolving CO_2 the rate of CO_2 hydration is of similar magnitude as physical mass transfer from gas into liquid phase [119].

3.2 Effect of Carbon Dioxide on Microbial Growth

It has been long known that CO_2 can inhibit microbial growth. This effect has been exploited for the preservation of packed food, e.g. meat. However, as is evident from Fig. 3, the influence of CO_2 partial pressure[1] on aerobic microbial growth is not uniform. There is a general tendency of growth retardation by CO_2 in the examples shown in Fig. 3; but in two cases [31,33], lower CO_2 partial pressure stimulated growth, and only at higher p_{CO_2} was growth inhibited.

In their investigation of a strain of *Ps. fluorescens* Gill and Tan [33] discovered the interesting result, that in complex media, growth rate decreased linearly with increasing CO_2 partial pressure, whereas in minimal media, low CO_2 partial pressure was growth promoting and higher p_{CO_2} (> approx. 200 mbar) growth retarding. It can be assumed that in the latter case, i.e. in a minimal medium, CO_2 was needed by the culture for the biosynthesis of essential compounds via carboxylation reactions. In a complex medium these essential compounds are already present. Thus in a minimal medium, growth inhibition by CO_2 was effective only at higher CO_2 concentrations. This latter result is in agreement with our investigation of *Ps. fluorescens* on a minimal medium [35], yielding significant growth retardation at CO_2 partial pressure of 200 and 400 mbar with growth rates reduced to 70% resp. 60%; in the range of low CO_2 partial pressures (up to 40 mbar) we could not observe a significant effect on growth rate.

---

[1] Usually the effect of CO_2 is expressed as being a function of the partial pressure of CO_2. Of course, the actual effect of CO_2 is caused by its concentration in the medium, which only at saturation, i.e. in equilibrium with the gas phase, is directly related to its partial pressure. However, in culture vessels with efficient exchange between gas and liquid phase saturation equilibrium will prevail approximately, i.e. within limits of experimental accuracy. Besides, in practically all published investigations on CO_2 effects, only CO_2 partial pressures have been measured, but not CO_2 concentrations. Therefore in Fig. 3 the partial pressure of CO_2 is used as the independent variable.

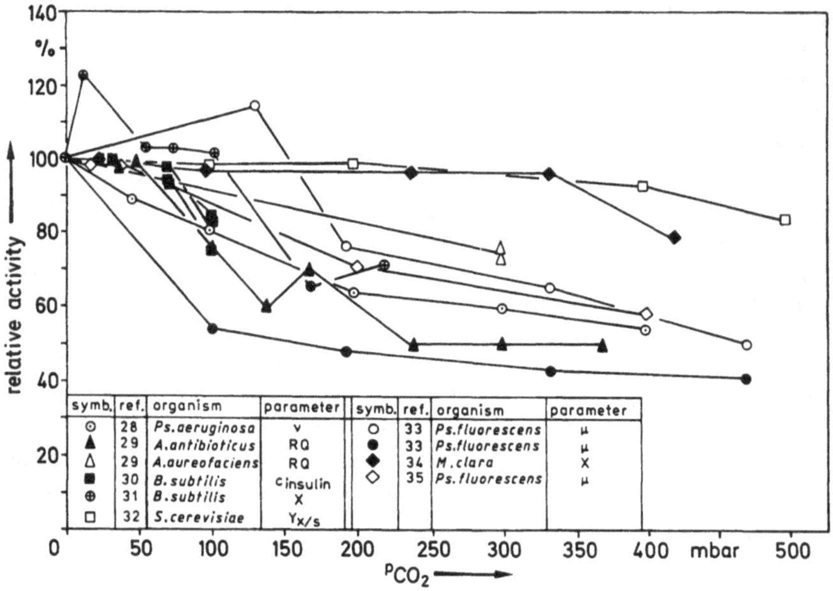

Fig. 3. Effect of carbon dioxide partial pressure p_{CO_2} on growth of heterotrophic microorganisms. Microbial activity is given by data for growth parameters related to values at 1 bar air pressure = 0,35 mbar CO_2; (*A.* = *Actinomyces*, *M.* = *Methylomonas*)

In our continuous cultivation experiments of the same strain of *Ps. fluorescens* at pressures of 1,15 to 8,0 bar yielding total inhibition of growth at 8,0 bar, we observed CO_2 partial pressures ranging from 10 to 32 mbar (based on measurement in the exhaust gas) [20, 21]. Obviously CO_2 was not responsible for growth inhibition in these experiments.

Similar results have been obtained by Ho and Smith [36] for penicillin production. They observed inhibition of mycelial growth and penicillin production only at CO_2 partial pressures of 126 mbar and above; at 30 and 50 mbar they found no effect. Pirt and Mancini [37] found a reduction in penicillin production of 50% at 80 mbar CO_2, and Bylinkina et al. [29] report a considerable decrease in respiration and antibiotics production for three strains of *Actinomycetes* at 150 to 200 mbar CO_2 saturation of the medium. Formation of inosine by *B. subtilis* is inhibited at a CO_2 partial pressure of more than 50 mbar [30]. On the other hand production of α-amylase by *B. subtilis* increased by a factor of three at 80 mbar CO_2 [31], and maximum formation of proteases has been observed at 90 mbar CO_2 [38]; though in both cases growth was inhibited. Finally an interesting result should be mentioned which demonstrates the variety of CO_2 effects on metabolism of microbes. In an investigation of the effects of increased pressure on the facultative autotrophic, thermophilic bacterium *Sulfolobus acidocaldarius* grown on a heterotrophic medium (yeast extract) aerated with air or a mixture of air and 5% (v/v) CO_2, Sturm et al. [39] found total growth inhibition at a CO_2 partial pressure of 250 mbar and more, though this chemolithotrophic bacterium is able to fix CO_2 with elemental sulfur as energy source. For their experiments the authors used a high pressure autoclave rated to 200 bar and 250 °C. By adding helium to the aerating.gas they performed cultivation experiments up to 120 bar at 67 to 85 °C. Generally, it can be stated that carbon dioxide has a dual effect on cell growth since low concentrations may be required for growth and high concentrations are growth inhibiting [40]. However, the degree of growth restriction by CO_2 may differ widely depending on the species of microorganism and on environmental parameters such as the composition of medium and substrate. Therefore a simple explanation of growth inhibition by CO_2 is not to be expected. Most probably mass action effects of dissolved CO_2 on decarboxylation play a role [41], but CO_2 inhibition of metabolic enzymes cannot be excluded aither [33].

Regarding possible effects of CO_2 on aerobic cultures in industrial bioreactors, we can start with an estimation of maximum possible partial pressures and liquid concentrations of CO_2. In air of 5 bar with 0.03 vol-% CO_2 partial pressure of CO_2 is 1.5 mbar. With production of CO_2 the off-gas from the bioreactor may contain 1 to 3 vol-% CO_2. This value will be reached at the top of the bioreactor with a total pressure of approx. 1.5 bar, corresponding to CO_2 partial pressures of 15 to 45 mbar. Taking into account a certain degree of supersaturation, CO_2 concentrations in the medium will amount to values corresponding to equilibrium partial pressures of perhaps 25 to 75 mbar. This means that in part of the reactor inhibitory effects of CO_2 may become marked, depending on species and state of the culture and on environmental conditions. On the basis of present knowledge these effects should not be large; in scale-up of processes they should, however, not be neglected.

4 Oxygen Partial Pressure

4.1 Dissolved Oxygen Concentration — Intrinsic Parameter

Molecular oxygen is needed by aerobic microorganisms as an oxidant to meet their requirements of energy. The sources of energy are carbon compounds, e.g. carbohydrates, alcohols, hydrocarbons, which are finally oxidized to carbon dioxide and water. This process, called respiration, takes place via various pathways of sequences of chemical reactions, depending on substrate and type of microorganism. The velocity of these reactions and consequently the rate of microbial growth will depend on the concentrations of reactants in the medium. This is also true for oxygen. However, in contrast to the other reactants, the concentration of oxygen is limited to low values because of its low solubility in aqueous culture media. At 25 °C the concentration of oxygen in water saturated with air of 1 bar is 0.263 mol m^{-3} (= 8.4 ppm). Similar to CO_2, solubility of oxygen is decreased by dissolved electrolytes and organic substrates except short-chain alcohols (for estimation of oxygen solubility in culture media see [25,26]).

To exemplify the consequences of the low solubility of oxygen on aerobic microbial growth let us consider a typical yeast culture on carbohydrates during exponential growth with specific growth rate $\mu = 0.5\ h^{-1}$ and oxygen demand of 1.5 g O_2 per g biomass. At 30 °C and 100% saturation in equilibrium with air of 1 bar the culture medium would contain 7.5 ppm oxygen. Assuming a cell mass concentration of 10 g l^{-1} this dissolved oxygen would be totally consumed by the culture within 3.6 s, if oxygen supply by aeration were suddenly stopped.

This example impressively demonstrates the importance of efficient mass transfer from aerating gas into the liquid medium. In order to promote mass transfer and to increase dissolved oxygen concentration, aeration with higher pressures or with oxygen-enriched air has often been proposed and in several cases been applied [1,6,42,43].

For gluconic acid production from glucose May et al. [44] and Wells et al. [45] found an increase in gluconic acid yield from approx. 30% to more than 80% by applying pressures of 6.0 bar in the first case and 3.2 bar in the second. In the two papers, no information on dissolved oxygen concentration is given; but considering mass transfer conditions in the experiments of both research groups — May et al. [44] used gas-washing bottles and Wells et al. [45] rotary drums as bioreactors — it can be assumed, that dissolved oxygen concentrations must have been rather low. This means that limitation of oxygen transfer had been reduced by applying higher pressures and consequently higher oxygen partial pressures. This conclusion is confirmed by a recent investigation with cultures of *Gluconobacter oxydans* on a synthetic medium [46] in which gluconic acid yield was more than 80% and was shown to be independent of oxygen partial pressure up to 2.0 bar. At the same time, rates of both microbial growth and product formation decreased linearly with oxygen partial pressure and also with dissolved oxygen concentration as the intrinsic parameter for microbial reactions; at 2.0 bar oxygen tension rates of growth and gluconic acid formation were practically zero (see Fig. 4). For the same process similar results were obtained by Oosterhuis et al. [47] culturing *G. oxydans* on a complex medium. For practical reasons, i.e. because of the difficulties in measuring actual dissolved oxygen concentrations, very

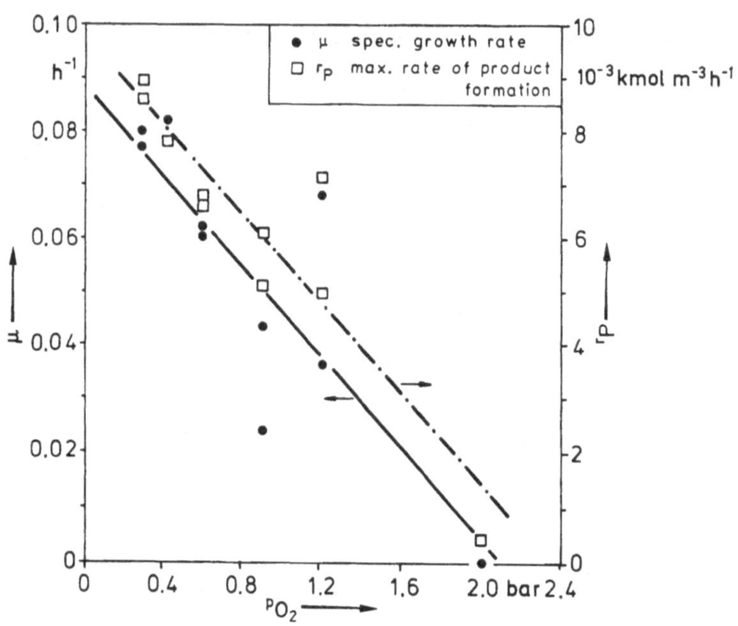

Fig. 4. Influence of oxygen partial pressure p_{O_2} on specific growth rate μ and rate of gluconic acid formation r_p in cultures of *Gluconobacter oxydans*; data from Träger et al. [46]

often in experimental work on effects of oxygen on microbial growth oxygen partial pressure in the gas phase is used as parameter, and implicitly 100% oxygen saturation of the liquid medium is assumed. This is not necessarily true, as shown above for some results published on gluconic acid production. Therefore data from the literature on effects of oxygen partial pressure on microbial growth have to be interpreted with care.

In this respect, another point has to be considered. Dissolved oxygen concentration is very often given as partial pressure of oxygen. This wording lacks clarity, since partial pressure has been defined originally for ideal gas mixtures:

$$p_i = y_i P \tag{6}$$

p_i partial pressure of component i
y_i molar fraction of component i in gas mixture
P total pressure

For liquid mixtures, such as nutrient media, the expression "partial pressure" is used to characterize the concentration of a volatile (or gaseous) component i in the liquid (L) by stating the concentration of i in a gas phase (G) in *equilibrium* with the liquid phase. More precisely the condition of phase equilibrium is equality of fugacities:

$$f_i^L = f_i^G \tag{7}$$

f_i^L, f_i^G fugacity of i in liquid (L), or gas (G) phase
Fugacities are "corrected pressures"; in ideal gas mixtures (and real gases up to several bar can be considered ideal) they are equal to the partial pressure p_i:

$$f_i^G = p_i \quad \text{for} \quad P < 5 \ldots 10 \text{ bar} \tag{8}$$

Oxygen concentrations in nutrient media are usually measured with the aid of electrochemical electrodes, which are calibrated by immersing them in a liquid medium equilibrated with a gas phase of known oxygen concentration (e.g. air) and pressure, that is at definite oxygen fugacity $f_{O_2}^G$ equal to oxygen partial pressure p_{O_2} [eq. (8)]. Because of this procedure unfortunately very often a so-called oxygen partial pressure p_{O_2} is used as a measure of dissolved oxygen concentration whereby oxygen saturation of the liquid at that partial pressure is tacitly assumed. To avoid misunderstandings, in this review we use the term "dissolved oxygen tension" for dissolved oxygen concentration when citing such data from literature references.

4.2 Analysis of Effects of Total Pressure and Oxygen Partial Pressure

4.2.1 Experimental Techniques

Investigations of effects of oxygen partial pressure on microbial cultures can be carried out in two ways. The experiments can be performed either at different total pressures with air or at constant total pressure using gas mixtures with different oxygen content for aeration, e.g. mixtures of air and oxygen. The latter way will of course yield unequivocal results as to the effects of oxygen partial pressure; it requires, however, a controlled gas mixing device producing gas mixtures with constant composition. This may be one of the reasons why many researchers have carried out their investigations with air, varying total pressure. This procedure is justified, when total pressure has only minor effects, if any, on microbial growth in the range in which total pressure is varied.

A large number of investigations on effects of oxygen partial pressure on microbial cultures has been performed by putting test tubes or petri dishes in incubators, desiccators or pressure chambers filled with pure oxygen or oxygen-air mixtures at 1 bar [48,49,50]; in some cases higher pressures have been applied in installations of this type [51]. This method of experimentation is well suited for investigations, in which the effect of oxygen on a multitude of strains of microorganisms is to be tested. Research of this kind has been initiated largely by possible applications of toxic effects of oxygen on infectious germs (see e.g. Gottlieb [52]). For investigations aimed at finding more quantitative effects of oxygen on individual strains of microorganisms cultivation in bioreactors is to be preferred, because here better control and monitoring of experimental parameters is possible. Experiments of this type in standard stirred bioreactors have been carried out by various authors, either aerating the bioreactor with pure oxygen or oxygen-air mixtures at 1 bar [34,53,54] or putting the bioreactor into a pressure chamber [55]. For several years, stirred bioreactors for higher pressures (up to approx. 10 bar) have been available (e.g. from Bioengineering AG, Wald/Switzerland, or Marubishi Bioengineering Co., Ltd., Tokyo/Japan).

The problem of sealing the stirrer shaft tight is avoided with airlift reactors. An apparatus of this type developed for bioprocesses at higher pressures [56] is shown in Fig. 5. It is an airlift reactor with internal circulation, made from stainless steel. Gas

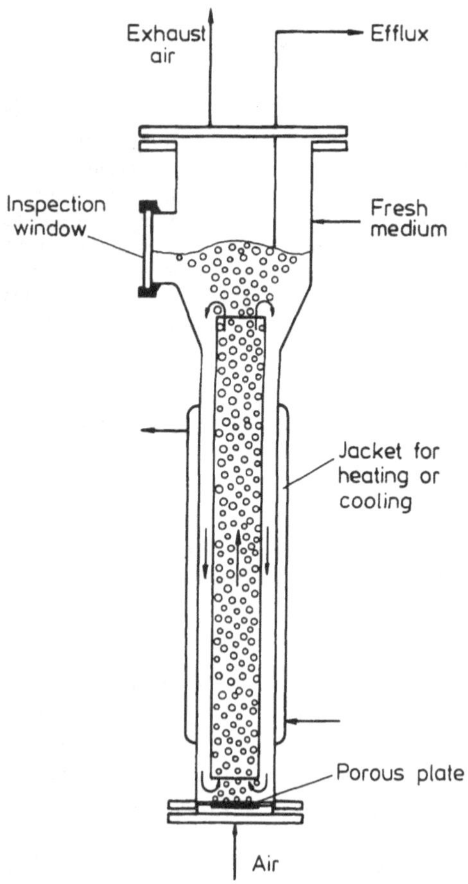

Fig. 5. Stainless steel airlift bioreactor for cultivation at higher pressure [56]. Total volume 20 l, working volume 10 l

for aeration is introduced below the inner tube via porous plate or perforated tray. The outer tube carries a jacket for heating or cooling. The upper end of the reactor is cone-shaped to promote degassing and defoaming. An inspection window here allows one to observe the level of the liquid medium and the efficiency of defoaming. The upper plate is equipped with ports for measuring probes, exhaust gas and withdrawal of medium under continuous operation. The bioreactor shown in Fig. 5 has a total volume of 19 l: for optimum operation it should be filled 50 to 55% with liquid medium. Maximum operating pressure is 10 bar; with the same design higher pressures are possible with correspondingly thicker outer walls. Circulation of the liquid content is rather fast with circulation times of about 5 s; this means, that in continuous operation the reactor is practically ideally mixed [57, 120].

4.2.2 Procedure and Analysis of Experiments

It seems that effects of oxygen on aerobic cultures will be detected unequivocally, when comparative experiments are performed, e.g. by aerating the medium after inoculation in one experiment with air and in a second experiment with oxygen. If one is lucky, results are obtained of the type shown in Fig. 6. Apparently on the complex medium *Ps. fluorescens* grows regardless of the oxygen content of the aerating gas, whereas on the synthetic medium, growth is totally inhibited under pure oxygen. This result may be explained by an inhibitory effect of oxygen on the culture with synthetic medium. But this is not the only possible explanation. As has been shown in Section 3.2, low concentrations of CO_2 may be necessary, which means that total lack of CO_2, as in pure oxygen, will retard growth. In fact, Gottlieb [48] found that growth

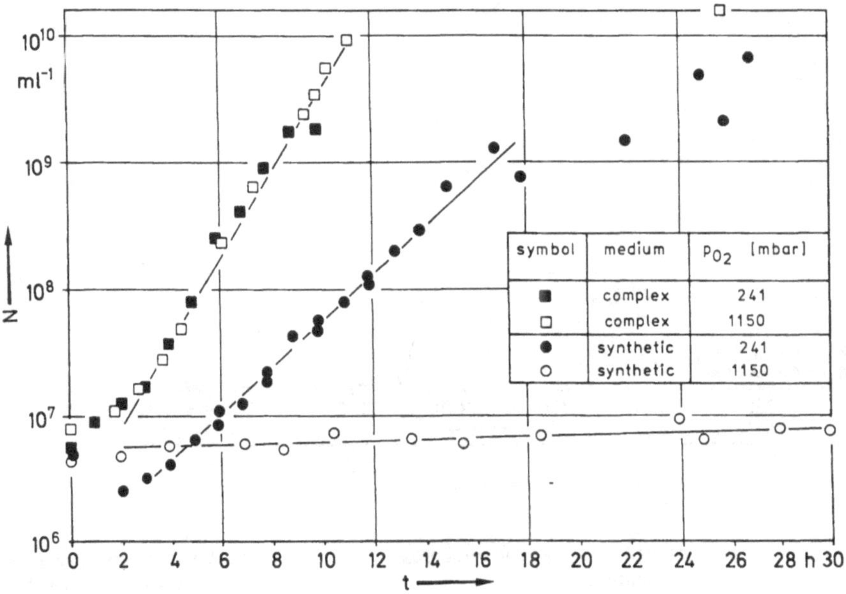

Fig. 6. Effect of oxygen partial pressure p_{O_2} and composition of nutrient medium on growth of *Pseudomonas fluorescens* [35]. Total pressure P = 1,15 bar, temperature T = 30 °C

inhibition of *Achromobacter* by pure oxygen (1 bar) was markedly reduced, when 0.05 vol-% CO_2 was added to oxygen. A few more cases of growth inhibition by lack of CO_2 in cultures with pure oxygen have been reported (see Gottlieb [52]).

A way to identify growth inhibition by oxygen is to set up cultivations with air at higher pressures whereby oxygen partial pressure is increased proportionally. Results of such experiments are, however, not fully unambiguous, since effects of total pressure and CO_2 partial pressure may interfere, though these influences are most probably of minor importance (see Sects. 2.2 and 3.2). As to the possible effects of CO_2, varying CO_2 concentration of the aerating gas in comparative experiments will yield the answer. To ascertain the effect of oxygen and to exclude eventual interference of total pressure, results from cultivations at different total pressure have to be compared with those, in which pure oxygen (and possibly oxygen-air mixtures) were used for aeration.

An example for this procedure is shown in Fig. 7 in which relative concentration of cell mass at continuous operation for *Ps. fluorescens* (synthetic medium) and *Methylomonas clara* have been plotted versus oxygen partial pressure for cultures aerated with air, pure oxygen or oxygen-air mixtures; a second scale gives the total pressure of air corresponding to the respective oxygen partial pressures. Obviously both cultures are inhibited by oxygen. Interestingly the range of oxygen partial pressure, in which oxygen inhibition becomes effective, is identical for the two cultures. This indicates

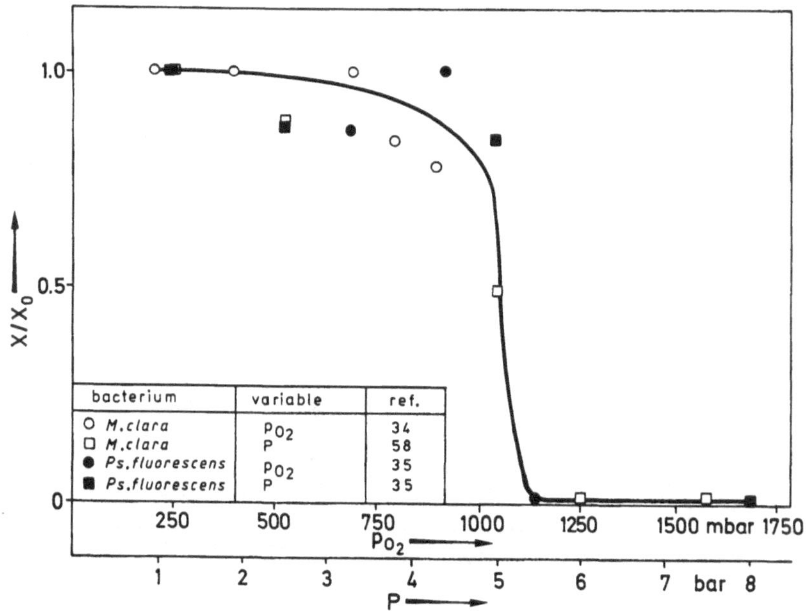

Fig. 7. Effect of oxygen partial pressure p_{O_2} and total pressure P on growth of *Pseudomonas fluorescens* and *Methylomonas clara* in continuous cultures (D = 0,15 ... 0,19 h^{-1}). As measure of microbial growth relative cell mass concentration $\frac{X}{X_0}$ is given with X_0 as cell mass concentration in cultures aerated with air at P = 1,15 bar ($\hat{=}$ p_{O_2} = 241 mbar).

that oxygen inhibition should originate from an identical cause in these two cases (see Sect. 4.3.3). Incidentally, CO_2 was not involved in either case [35, 58].

4.3 Effect of Oxygen on Microbial Cultures

4.3.1 Growth

As any other substrate oxygen may be the limiting factor for growth in aerobic cultures, when its concentration in the medium falls below a critical value. This critical oxygen concentration depends not only on the strain of microorganism, but also on conditions and state of the culture [59, 60]; as a rule it will be in the range of 0.1 to 20% oxygen saturation with air at 1 bar, i.e. approx. 0.008 to 1.5 ppm. Oxygen concentrations as low as that will occur in aerobic submerged cultures with efficient aeration, when the medium becomes highly viscous, as is the case in microbial synthesis of polysaccharides and of several antibiotics, in the latter case primarily caused by mycelial growth. The fact, that higher oxygen concentrations, i.e. from about 100% saturation with oxygen at 1 bar onwards, may have deleterious effects on cultures of microorganisms has been known for a long time. In 1909 and 1910 Moore and Williams [61, 62] published two investigations on the influence of an atmosphere containing 70 vol.-% oxygen on 26 species of infectious microorganisms. They had found that for two of these, *Bacillus tuberculosis* and *Bacillus pestis*, growth stopped completely; two species of *Staphylococci* grew more slowly than was normal, whereas growth of the other 22 species was not inhibited. In a subsequent investigation Adams [63] confirmed the results for *B. tuberculosis* and *B. pestis* at approx. 0.9 bar oxygen; an application of this effect for curing tuberculosis proved to be not feasible, because in experiments with guinea-pigs, prolonged treatment with an atmosphere containing more than 70 vol.-% oxygen led to lethal lung damage.

Since then a large number of investigations has been carried out on the effects of oxygen on microbial growth; a review by Gottlieb in 1971 [52] lists 137 species of microorganisms and 201 references. However, because of methodical deficiencies in many papers it is often rather difficult to draw unambiguous conclusions and to obtain quantitative relationships on the effect of oxygen.

Figure 8 gives such information on microbial growth as the ratio to data at 1 bar air ($\hat{=}$ 210 mbar O_2). As independent variable for the oxygen fugacity in the medium (cf. Sect. 4.1), the dissolved oxygen tension (DOT = oxygen "partial pressure") was chosen, because in most publications either this latter variable or relative oxygen saturation (ratio of oxygen content to saturation at 1 bar air or oxygen) is used. Figure 8 contains growth data only from such publications in which dissolved oxygen concentration or relative oxygen saturation of the medium has been measured or where practically 100% oxygen saturation at the actual partial pressure can be safely assumed from experimental conditions.

---

[2] From the two examples in Fig. 7 and from results in Fig. 8 and Table 3 it is evident, that the original cause for effects of total pressure on aerobic cultures up to 10 ... 60 bar given in Table 1 is oxygen partial pressure. This leads to the conclusion, that in this pressure range the effect of mere pressure is neglegibly small.

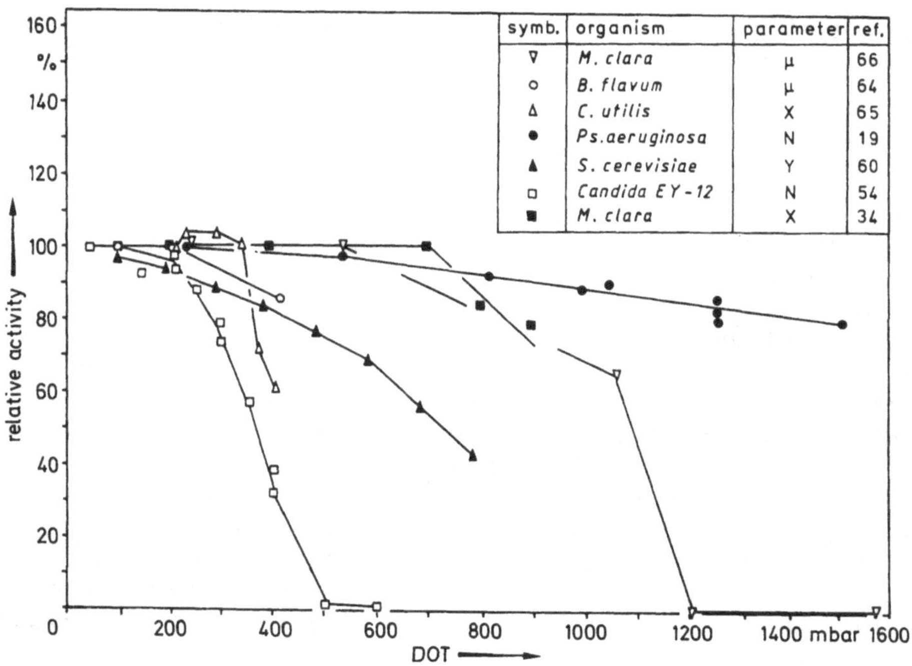

Fig. 8. Effect of dissolved oxygen tension (*DOT*) on microbial growth. Microbial activity is given by data for growth parameters related to resp. values at $DOT = 210$ mbar (saturation at 1 bar air pressure); (*B.* = *Brevibacterium*)

As can be seen from Fig. 8, the effect of oxygen on aerobic cultures is by no means uniform. Of course, conditions of cultivation, e.g. medium and substrate, were quite different in different investigations. Moreover, different parameters have been used in publications to represent oxygen effects. When several resulting parameters were given, we selected the one showing the most distinct effect.

More results on oxygen effects are summarized in Table 3. Notwithstanding strong differences in the influence of dissolved oxygen on microbial activity in aerobic cultures, it is evident from Fig. 8 and Table 3, that there exists a general tendency of inhibitory effects by oxygen. In some cultures (e.g. [54, 60, 65]) such effects become apparent at less than 400 mbar dissolved oxygen tension (corresponding to oxygen saturation with air at less than 2 bar), in others [34, 35] at about 1000 mbar dissolved oxygen tension. Interestingly for two species of the same family (*Pseudomonas*) distinct differences in the effect of oxygen have been reported (cf. [19] and [35] in Fig. 8), and Brown et al. [49] even found in an investigation of three strains belonging to the same species (*Ps. aeruginosa*), that one strain was sensitive to oxygen and the other two not. Differences in inhibition by oxygen were also observed between a wild strain and a production strain of *Str. aureofaciens* [66]. In view of this situation it is evident, that oxygen effects cannot be correlated with the taxonomy of microorganisms.

Table 3. Influence of Increased Oxygen Partial Pressure on Growth of Aerobic Cultures

Organism	Maximum oxygen partial pressure $p_{O_2 max}$ (mbar)	Mode of operation[a]	Effect	Refs.
Baker's Yeast	810	C	Decrease in cell yield by 60%	60)
Candida utilis	410	C	Maximum cell yield at p_{O_2} = 350 mbar	65)
Candida EY-12	1000	B	Total inhibition of growth at p_{O_2} = 500 mbar	54)
Saccharomyces cerevisiae	1000	C	Decrease in cell density by 25—40% depending on dilution rate	68)
Pseudomonas AM1	750	C	Decrease in cell yield by 27%	53)
Pseudomonas fluorescens	1670 (P = 8 bar)	B	Growth inhibition at p_{O_2} = approx. 1000 mbar, total inhibition at p_{O_2} = 1670 mbar	35)
Methylomonas clara	900	C	Decrease in cell yield by 21%	34)
Methylomonas clara	1200	B	Total inhibition of growth	66)
Micromonospora vulgaris	1000	B	Inhibition of respiration	67)
Micromonospora purpurea	1200	B	Total inhibition of growth	66)
Sulfolobus acidocaldarius	1050 (total pressure 50 bar)	B	Total inhibition of growth	39)

[a] B batch culture
C continuous culture

4.3.2 Formation of Products

The predominant interest in industrial processes is yield and rate of product formation. As regards the effects of increased oxygen concentration on the formation of products in aerobic microbial processes, not much has been published.

For citric acid formation by submerged batch cultures of *Aspergillus niger* on molasses, Clark and Lentz [69] report maximum production rate at an oxygen partial pressure of 1.7 bar, without giving information on dissolved oxygen concentration. From operating conditions of their experiments (maximum specific aeration rate at 0.2 vol. $vol.^{-1}$ min^{-1}, tower reactor) it must be assumed that mass transfer limitation was prevailing. This is supported by a study of citric acid formation with *A. niger* of Khan and Ghose [70]; they found, that citric acid yield increases with aeration effi-

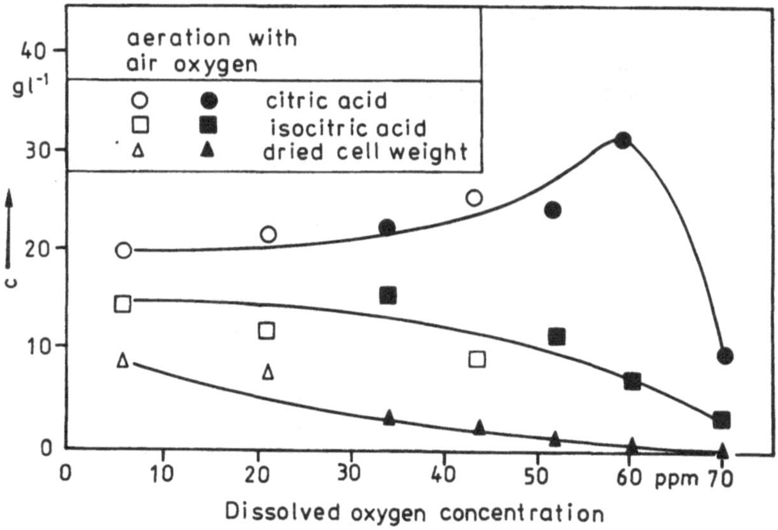

Fig. 9. Influence of dissolved oxygen concentration on microbial growth and product formation in cultures of *Candida tropicalis* [72]

ciency, whereas cell growth is not much affected. In continuous cultures of *A. foetidus* producing citric acid from glucose Kristiansen and Sinclair [71] found that citric acid concentration could be increased by approx. 25% by increasing dissolved oxygen tension from 120 to 300 mbar without change of cell mass concentration. Still higher yields of citric acid were obtained by Okoshi et al. [72] in batch cultures of *C. tropicalis* using air and oxygen with oxygen partial pressure up to 2000 mbar. These authors found optimum citric acid production at a dissolved oxygen concentration of 60 ppm (= 1700 mbar dissolved oxygen tension); at this optimum the concentrations of cell mass and isocitric acid were distinctly lower than with air at 1 bar (see Fig. 9). Summing up these results from different authors, it can be safely concluded that optimum dissolved oxygen tension for citric acid production is well above 1 bar, independent of strain of organism.

With gluconic acid production the situation is quite different. Here, we showed recently that in cultures of *Gluconobacter oxydans* on glucose the rates of cell growth and of product formation decreased in parallel linearly with increasing dissolved oxygen concentration [46], see Sect. 4.1). For the production of inosine by cultures of *B. subtilis*, Shibai et al. [73] clearly demonstrated, that inhibition of inosine formation due to insufficient oxygen supply can be avoided by aeration with oxygen-enriched air. Researchers from the same laboratory showed that L-glutamic acid production by cultures of *Corynebacterium hydrocarboclastus* on hydrocarbons can be enhanced by improving oxygen supply using oxygen-air mixtures [42]. Likewise Yamada et al. [74] assume increased oxygen transfer as the cause for the acceleration of sorbitol conversion to sorbose in shaken flask cultures of *Acetobacter suboxydans* by raising oxygen content in the gas phase from 20 vol.-% (air) to 40 vol.-%. Buckland et al. [75] found maximum production of cholesterol oxidase by *Nocardia rhodocrous* at 30 to 40% oxygen saturation with air at 1 bar.

Table 4. Influence of oxygen partial pressure on product formation

Organism	Product	Maximum oxygen partial pressure $p_{O_2\,max}$ (mbar)	Mode of operation[a]	Effect	Ref.
Acetobacter suboxydans	Sorbose (by oxidation of sorbitol)	810	B	Decrease in sorbose formation by 75% with air-grown cells as inoculum; with cells grown at p_{O_2} = 400 mbar increase in sorbose formation	74)
Gluconobacter oxydans	Gluconic acid	800	B complex medium	Decrease in rate of product formation by 85%	47)
Gluconobacter oxydans	Gluconic acid	2000	B synthetic medium	Decrease in rate of product formation to zero at $p_{O_2\,max}$.	46)
Aspergillus niger	Citric acid	2400	B	Maximum yield and rate of product formation at p_{O_2} = 1700 mbar	69)
Aspergillus foetidus	Citric acid	300	C	Increase in yield by 25%; no effect on cell density	71)
Candida tropicalis	Citric acid	2000	B	Maximum yield at DOT = 1700 mbar with decreased cell density	72)
Nocardia rhodocrous	Cholesterol oxidase	210	B	Maximum production at DOT = 65 ... 85 mbar; enhancement of selectivity by increasing DOT	75
Streptomyces aureofaciens	Tetracyclin	1670	B	Increase in product formation per cell mass; increased formation of oxytetracyclin	66)
Streptomyces rimosus	Oxytetracyclin	1200	B		66)
Micromonospora purpurea	Gentamicin	1200	B	Total inhibition of product formation	66)

[a] B batch culture;
C continuous culture

A survey of published data on effects of increased oxygen partial pressure on aerobic cultures with product formation is given in Table 4. From these findings it is evident, that elevated oxygen partial pressures may have distinctly different effects. Positive effects may be directly related to enhanced oxygen transfer, or they may be due to action on cell metabolism, whereas negative effects can only be caused by an influence on metabolism. Accordingly general rules on possible oxygen effects in microbial processes with product formation cannot be given at present; when such knowledge is required for a specific process, it can only be obtained from experiments.

4.3.3 Possible Causes of Toxic Oxygen Effects on Growth

It has been known for a long time that oxygen may have toxic effects on aerobic microorganisms under oxygen partial pressures not much higher than in air at 1 bar. This seems to be surprising at first sight, because aerobes need oxygen for their metabolism. However, in the evolution of life on earth anaerobic metabolic processes were developed first at a time when the atmosphere contained practically no oxygen. Only when oxygen had been generated by photosynthesis in appreciable quantities, could aerobic microorganisms develop. Therefore it is generally accepted now, that oxygen is toxic at any concentration, and that aerobic organisms have developed mechanisms to protect them against oxygen toxicity. When these mechanisms are disturbed by higher oxygen concentrations, toxic effects will appear [76,77,78].

As to the way in which inhibition of aerobic microorganisms by oxygen proceeds, it should be noted, that the organisms are not killed immediately upon increase of oxygen partial pressure. As a consequence of higher oxygen concentration, the cells may retard or stop growth, depending on their sensitivity to oxygen and on actual dissolved oxygen concentration. For example, Brunker and Brown [55] found an increase in generation time by 33% for a strain of *E. coli* at 1 bar of oxygen compared to cultivation with air at 1 bar, whereas oxygen above 1,5 bar was completely bacteriostatic. Cells are usually still viable after exposure to increased oxygen partial pressure; they start to grow again on reincubation with air at 1 bar after a time lag which in some cases was found to be proportional to the duration of oxygen exposure [35,66,80]. Prolonged application of increased partial pressure of oxygen will eventually kill the cells. Duration of oxygen treatment after which cells are completely killed depends on species and oxygen pressure; it ranges from less than 24 h to more than 28 d [50,80,81,82]. These observations lead to the conclusion, that inhibition of metabolism by oxygen is connected with gradually proceeding damage to cells.

The ways which have been discussed in literature [76,78,83,84,85] that may be the possible causes of oxygen toxicity can be classified into three groups; either oxygen may directly affect certain enzymes; or it may form activated products, i.e. radicals and reactive molecules, which will attack and destroy biomolecules; or lastly, oxygen may shift the redox equilibrium inside the cell and disturb biosynthetic reactions by binding reductive equivalents. That oxygen can inhibit or inactivate enzymes, has been shown by several authors. For example Brown et al. [79] found that in *E. coli* quinolinate phosphoribosyl transferase which is essential for the biosynthesis of NAD, is inactivated at 4,2 bar oxygen, and Brown and Yein [86] identified dihydroxyacid dehydratase as the specific point of oxygen inhibition in biosynthesis of branched-chain amino acids. This explains the findings of Boehme et al. [87], that addition of certain amino acids to the medium can prevent or alleviate oxygen inhibition. Matsumura and Kobayashi [54] showed, that the decrease of specific growth rate of *Candida* sp. on ethanol with higher dissolved oxygen concentration is parallel to the decrease in activity of acetyl-CoA synthetase and isocitrate lyase. By varying the carbon source for *Ps. fluorescens*, Liefke [66] was able to attribute oxygen inhibition of these cultures to blockage of glyceraldehyde-3-P-dehydrogenase. It is known, that this enzyme is readily inactivated at elevated oxygen concentrations [88], presumably by oxidation of thiol (-SH) groups to disulphides [76, 89]. Damage to protein structure by this reaction may be the cause of deactivation by oxygen for many other enzymes, too [83]. Thiol oxidation may lead to intramolecular or intermolecular disulfide bonding. As a consequence the tertiary structure of proteins will be altered, and prosthetic groups may dissociate. Besides such an irreversible inactivation by oxygen, reversible inhibition of enzymes is possible as was shown for glyceraldehyde-3P-dehydrogenase of *Ps. fluorescens* [66]. Two mechanisms may be responsible for the reversible inhibition of glyceraldehyde-3P-dehydrogenase. Either oxygen acts as a competitive inhibitor, as was indicated at dissolved oxygen concentrations up to 25 mg l^{-1} [35], or oxidized thiol groups are reduced by reducing agents, e.g. glutathione, thus reactivating the enzyme.

Mechanisms for the toxic effects of oxygen are not yet clear. That the O_2 molecule (dioxygen) itself may cause cell damage is not probable, since the activation energy is rather high [90]. However, components of the cellular substance may act as electrone donors, generating partially reduced oxygen species of high reactivity, e.g. superoxide radical ($O_2^{\cdot -}$) or the peroxide ion (O_2^{2-}); from these molecules the extremely reactive $HO^{\cdot}$ radical is formed in the iron-catalyzed Haber-Weiss reaction (cf. Table 5) [90–94]. The superoxide radical itself is not very reactive in aqueous solution, but its reactivity is increased manyfold in a hydrophobic environment, e.g. in some regions of the cellular membrane [94].

Besides oxygen-derived radicals and peroxides singlet oxygen ($^1O_2$) [95] and long-lived secondary radicals which are formed by the reaction of $HO^{\cdot}$ or $O_2^{\cdot -}$ with amino acids, nucleotides and especially phospholipids are possible causes of the toxic effects of oxygen [96]. Bruyninckx et al. [97] and Farr et al. [98] demonstrated that oxygen radicals

Table 5. Main reactions of univalent oxygen reduction

$O_2 + e^- \longrightarrow O_2^-$	(9)
$O_2^- + O_2^- + 2\,H^+ \xrightarrow{SOD} O_2 + H_2O_2$	(10)
$Fe^{3+} + O_2^- \rightarrow Fe^{2+} + {}^1O_2$	(11)
$Fe^{2+} + H_2O_2 + H^+ \rightarrow Fe^{3+} + HO^{\cdot} + H_2O$	(12)
$H_2O_2 + O_2^- + H^+ \rightarrow HO^{\cdot} + {}^1O_2 + H_2O$	(13)
$H_2O_2 + H_2O_2 \xrightarrow{Catalase} O_2 + 2\,H_2O$	(14)
$RH_2 + H_2O_2 \xrightarrow{Peroxidase} R + 2\,H_2O$	(15)

SOD: Superoxide dismutase;
Eq. (11): Fenton reaction;
Eq. (13): Haber-Weiss reaction

even cause mutagenic effects. Environmental conditions, e.g. pH and certain transition metals such as Cu, Fe, markedly affect oxygen toxicity. By protonation of the superoxide radical the perhydroxyl radical HOO˙ is formed which shows considerably higher reactivity towards biomolecules [99].

Although it is not possible to present definite mechanisms for oxygen toxicity and to decide which reactive oxygen species causes deleterious effects on intracellular compounds, there is no doubt, that all the reactions mentioned so far occur in living cells. It is generally accepted, that oxygen toxicity is initiated by univalent reduction of the O_2 molecule which finally leads some way or other to damage of enzymes, nucleic acids or lipids. As a safeguard against reactive species generated by oxygen, aerobic organisms have developed enzymes which are able to transform these reactive species into non-reactive ones.

The main reactions of univalent oxygen reduction and protective enzymes are listed in Table 5, according to which superoxide dismutase, catalase and peroxidase are the most important protective enzymes in the cell. Besides these enzymes the cell contains a large amount (up to 0,1 mMol) of antioxidants such as glutathione and ascorbate to protect itself against a sudden increase of dissolved oxygen concentration.

Apparently, aerobic microorganisms are able to adapt their enzymatic activity to elevated oxygen concentrations, though of course not unlimited. For the oxidation of sorbitol to sorbose by *Acetobacter suboxydans* an increase of oxygen partial pressure above 400 mbar caused inhibition of cell growth and sorbitol conversion, when the inoculum had been cultivated under air at 1 bar ($\triangleq$ 210 mbar O_2); but when the inoculum had been grown at 400 mbar O_2, rates of growth and of sorbitol conversion at 800 mbar O_2 were still higher than at 400 mbar O_2 [74]. Lee and Hassan [68] studied the influence of increased oxygen partial pressure on chemostat cultures of *Saccharomyces cerevisiae*. The exposure of the cells to 100% O_2 induced superoxide dismutase and repressed formation of catalase and cell mass. Compared to air-grown cells, in pure oxygen the specific superoxide dismutase activity was 80 to 160% higher depending on dilution rate, whereas the maximum specific catalase activity was depressed to about 6% of the original value. These changes in enzyme activity were accompanied by a 25 to 40% decrease in cell mass concentration.

But increased concentrations of protective enzymes are not the only mechanisms, by which cells act against toxic effects of elevated oxygen concentration. Thus several authors [48,56,100] observed greater rates of respiration at higher oxygen concentrations. A very similar mechanism is known from obligate aerobic *Azotobacteraceae* containing oxygen sensitive nitrogenase. In order to protect this nitrogenase against oxygen, these bacteria either dramatically increase respiration of they reversibly switch off their nitrogenase activity [101,102]. Also changes in conformation of enzymes, i.e. different steric arrangements, have been suggested as protective mechanisms against oxygen exposure.

Summing up the broad variety of facts and conclusions, it is obvious that no simple explanation for the action of oxygen on aerobic cells can be given. This is not surprising, when it is recalled, that living cells are highly developed and complex systems whose structure is not static, but changing with time. The dynamic character of living cells is relevant to the following paragraph.

5 Timed Variations of Oxygen Partial Pressure

5.1 Experimental Techniques

Experimental investigations of the response of systems to changes of an external variable can be performed by applying single (step or pulse) or cyclic variations of the variable onto the system. In both cases, with oxygen partial pressure as the input parameter for an aerobic culture, two methods are possible: either a gas of constant composition, i.e. air, is used for aeration and total pressure is varied, or total pressure is kept constant and composition of aerating gas is changed, e.g. by changing mixing ratio of oxygen-air mixtures.

For the object of simulating the effects of local changes of environmental conditions on microbial cultures in industrial bioreactors the first method, i.e. varying total pressure, seems to be more realistic; but it bears the practical problem, that lowering total pressure will enhance foaming. Vardar and Lilly [103], who employed the method on penicillin production, report excessive foaming of their culture particularly at the venting stage, when they applied sinusoidal oscillations of total pressure with periods of 2 min. Kataoka et al. [104] apparently did not have such difficulties in their investi-

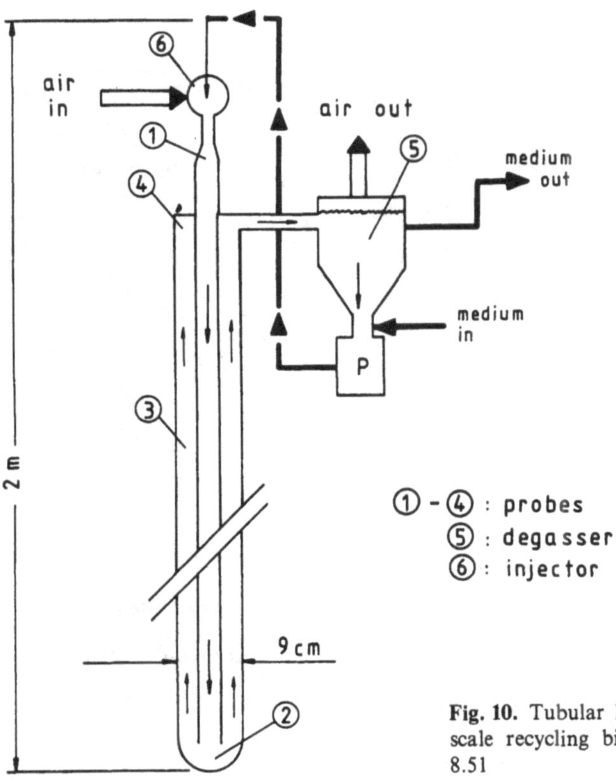

Fig. 10. Tubular loop reactor for simulating large scale recycling bioreactors [107]. Working volume: 8.5 l

gation of *Ps. aeruginosa* with oscillating pressure. Both research groups used a system of control valves for generating the pressure oscillations, whereas Takamatsu et al. [105] employed a moving piston installed at the top of a stirred tank reactor in order to simulate a deep-tank process for aerobic waste water treatment. For the second method of applying oscillating oxygen pressures — changing the oxygen content of the aerating gas — Soni and Ghose [106] described a simplified experimental set-up in which the aerating gas is supplied alternately from two gas cylinders containing gases at different oxygen content. They used this arrangement for an investigation of growth dynamics of *Saccharomyces cerevisiae* by aerating the culture alternately with pulses of pure oxygen and nitrogen.

An installation designed for the purpose of simulating large-scale recycle bioreactors has been described by Katinger [107] (see Fig. 10). It is a tubular loop reactor with draught tube as downcomer and annular space as riser; aeration is achieved by means of an injector located at the top of the downcomer. Recycle times are about 80 s, mixing times in the order of 5 min. With cultures of fairly high oxygen demand very steep gradients of dissolved oxygen concentration can be realized in this apparatus.

Process behaviour in response to variations of operating parameters is preferably studied in continuous operation. In the case of microbial processes special precautions and procedures are necessary to keep the culture sterile. An installation designed for

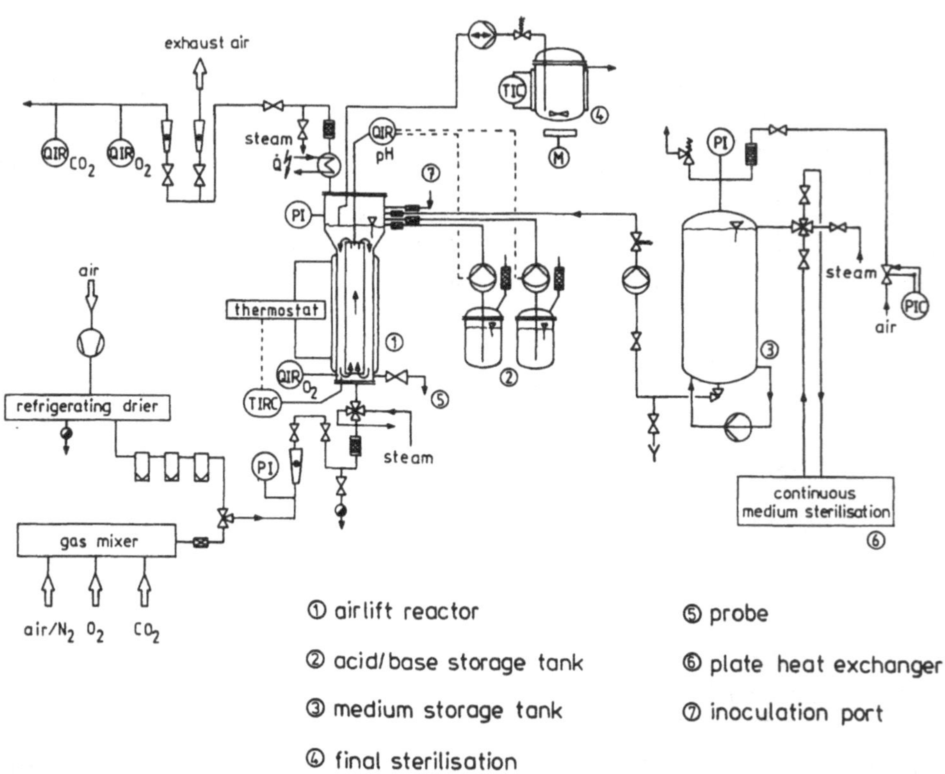

Fig. 11. Experimental setup for continuous cultivation [56]

this purpose is shown in Fig. 11. It is equipped with control instruments for temperature, pressure, pH, and defoaming. The medium is fed from a vessel where it has been stored after sterilization. Careful testing and skilled operation are required to run such units for longer periods without interruption.

5.2 Effects of Single Shifts

In several investigations it has been found that aerobic cultures, after being exposed to inhibitory oxygen partial pressures, may react with adaptive responses [74,108,109]. Especially interesting are a few cases, where cultures, which had first been subjected to increased oxygen partial pressures and subsequently to air at 1 bar, showed distinctly higher growth rates in comparison to cultures permanently grown at normal conditions, i.e. under air at 1 bar.

MacLennan et al. [53] report, that in a continuous culture of a strain of *Pseudomonas* on methanol, carbon-to-cell conversion was decreased, when oxygen partial pressure was raised stepwise from 130 to 850 mbar; however, when oxygen partial pressure was lowered to its initial value of 130 mbar, carbon-to-cell conversion increased again, but did not attain its original value. As possible causes, genotypic or phenotypic changes have been assumed by the authors. For reaching steady state 2 to 3 d (7 to 10 generations) were necessary. Here also the results of Yamada et al. [74] have to be mentioned; they found in shaken flask cultures of *Acetobacter suboxydans* that inocula from cultures grown in an oxygen-enriched atmosphere (40 and 80 vol.-% O_2) did not show growth inhibition when cultivated under higher oxygen partial pressure (up to 800 mbar O_2) in contrast to inocula grown under air, whose growth was markedly retarded when exposed to an oxygen-enriched atmosphere.

In continuous cultures of *C. utilis* on saccharose with air, pressure shifts from 1,5 to 7,0 bar (≙ 310 to 1460 mbar O_2) produced a 20% decrease and those from 7,0 to 1,5 bar a 20% increase in specific growth rate; but very quickly, i.e. within 1 to 2 generations, growth rates returned to their initial value [20]. The culture medium was saturated practically 100% with oxygen at the resp. partial pressure at the conditions of these experiments. This was not the case in the work of Páca and Grégr [65,110], who performed the continuous cultivation of *C. utilis* on ethanol in a four-stage tower bioreactor. Aerating the culture with atmospheric air at 1 bar led to oxygen limitation in parts of the bioreactor (third and fourth stage). By using a mixture of air and oxygen (35 vol.-% O_2) this limitation was avoided, and productivity increased considerably with the consequence, that biomass concentration in the product stream was doubled. But what is still more interesting: by changing the oxygen concentration in the aeration gas periodically, increase in biomass productivity was as high as with constant oxygen enrichment in the aeration gas. Optimum periods were 24 h with 35 vol.-% oxygen followed by approx. 50 h with air; only after this period did biomass concentration start to decrease, if oxygen content in the aeration gas was not raised again [110].

Similar long-term effects of increased productivity have been found by us with cultures of *Ps. fluorescens* on glycerol [35] and *Methylomonas clara* on methanol [58]. Both cultures show oxygen inhibition with complete stoppage of growth at oxygen partial pressures of more than about 1.2 bar (cf. Sect. 4.2).

In continuous cultivation, however, both species exhibit remarkable behaviour. Under permanently higher oxygen partial pressure, growth inhibition takes place, as

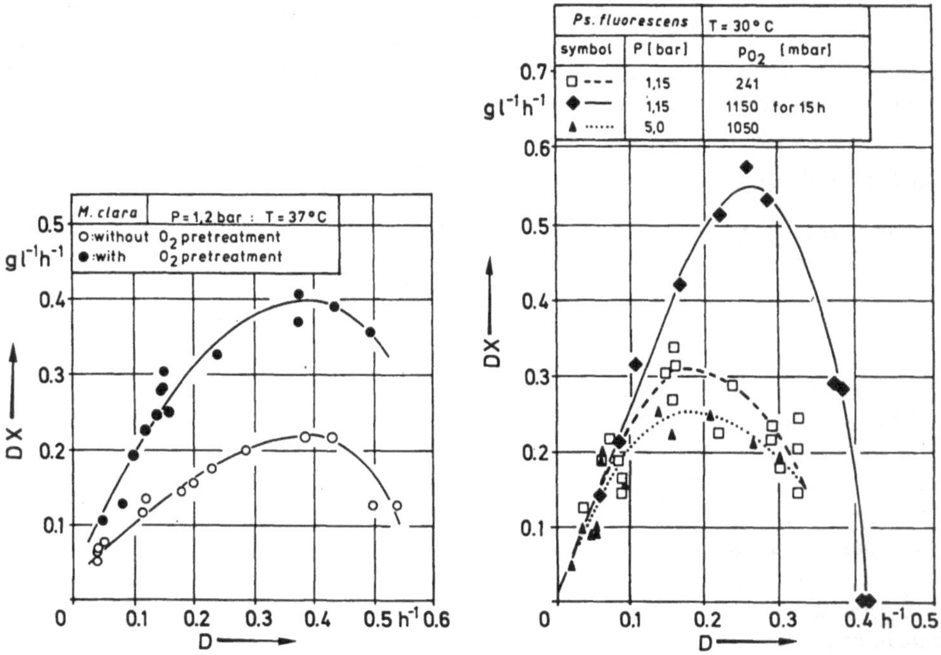

Fig. 12. Increase of productivity (*DX*) in continuous cultures of *Pseudomonas fluerescens* [35] and *Methylomonas clara* [58] after pretreatment with pure oxygen during discontinuous start-up phase. After inoculation the cultures were aerated with pure oxygen for 15 h and subsequently with air

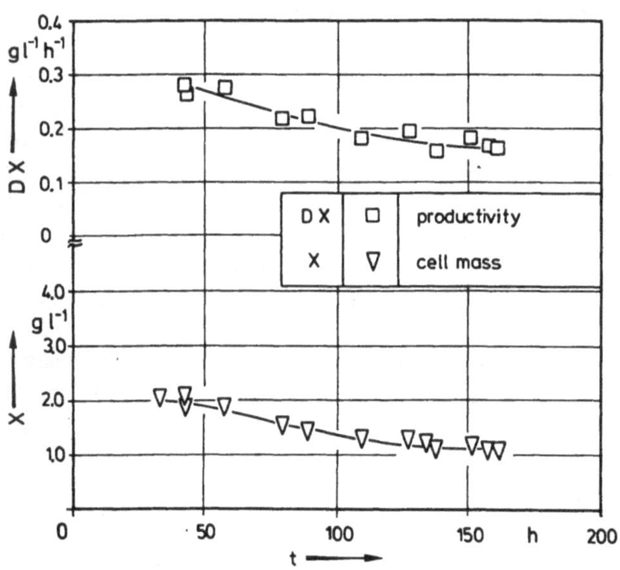

Fig. 13. Long-time decline of productivity increase of cultures of *Pseudomonas fluorescens* pretreated with pure oxygen (cf. Table 12) [35]; $D = 0{,}15\ h^{-1}$

is to be expected from the results of batch cultivations. But when, after the initial phase of batch cultivation, oxygen partial pressure is reduced by changing aeration to air at approx. 1 bar, productivity increases to distinctly higher values compared to cultures aerated with air from the beginning, i.e. from starting the discontinuous cultivation phase by inoculating the medium (see Fig. 12). Increase in productivity especially large at higher dilution rates where cell density is nearly doubled. These enhancements gradually decrease; after approx. 100 h productivities and cell mass concentrations have dropped to normal values, i.e. to values of cultures permanently aerated with air at 1 bar (see Fig. 13). Obviously it should be possible to utilize this effect for the purpose of increasing productivity by way of periodic variation of oxygen content in the aeration gas, e.g. by periodic admixing oxygen with air.

An interesting effect in a quite different range of oxygen supply to microbial cultures has been reported by Harrison et al. [59,111,112]. In chemostat cultures of the facultatively anaerobic bacterium *Klebsiella aerogenes* they observed oscillations in dissolved oxygen with a period of 2 to 4 min, when dissolved oxygen tension dropped below a critical value in the range between 2,5 and 13 mbar, i.e. between anaerobic and aerobic metabolism. By monitoring NADH fluorescence, Harrison et al. [59,112] were able to ascertain, that NADH was oscillating with the same frequency as dissolved oxygen. These results demonstrate, that ingenious and detailed experiments are required to achieve a better understanding of the mechanisms of oxygen effects on microbial metabolism.

5.3 Effects of Cyclic Variations

In order to simulate the cyclic changes of environmental parameters, especially of dissolved oxygen concentration, taking place in industrial bioreactors, a few different experimental installations have been employed. Gow et al. [1] reported on experiments which had been performed for the development of the ICI process for single cell protein from methanol. In the set-up used in these experiments, permanent cycling of pressure and of concentrations of oxygen, carbon dioxide and methanol could be established; a description of this bioreactor has not been given. Dissolved oxygen tension was oscillated between zero and several hundred mbar with 665 mbar as the maximum. Above 535 mbar O_2 as upper limit, carbon-to-cell conversion decreased, whereas the specific growth rate was not affected by the oscillations.

Katinger [107] was the first to describe in detail a loop reactor especially designed for the simulation of periodically changing environmental conditions (cf. Sect. 5.1., Fig. 10). As the test system, Katinger used the continuous cultivation of *Candida tropicalis* on *n*-alkanes. With dissolved oxygen tensions of 75 to 130 mbar and 5 to 20 mbar at the positions of maximum and minimum dissolved oxygen concentration in the reactor, rather steep oxygen gradients were imposed on the system. During continuous cultivation, two types of oscillations in respiratory activity were observed with periodicities of 15 to 20 min and 3 to 8 min; the latter coincided with actual ATP content of the culture. It was not possible, however, to identify the causes of these two types of oscillations. Because of similarity with the observations of Harrison et al. [112,113] in homogeneous cultures of *Klebsiella aerogenes* (cf. Sect. 5.2), similar causes are obvious, but identification from the nearly inexhaustible list of possible reasons is impossible without further experimental facts.

Sokolov et al. [114] investigated the effect of cyclic changes in dissolved oxygen and pH on three yeasts (*C. utilis* on ethanol, *S. cerevisiae* and *C. scottii* on glucose). Dissolved oxygen was varied by changing aeration between air and nitrogen for periods ranging from 0.2 to 60 min. In general, the anaerobic period (gassing with nitrogen) effected growth negatively with increasing length; in one case (*S. cerevisiae*) short anaerobic periods (2 min) had no effect. In a subsequent study Lirova, Sokolov et al. [115] tried to obtain information on the morphological state of cells in chemostat cultures of *C. utilis*, when oxygen supply and pH were varied periodically, and to correlate these cyclic changes of operating parameters with morphological changes of the cells; but their findings did not yield detailed conclusions on mechanisms of growth behaviour.

In an investigation of aerobic effluent treatment in tower reactors with *Ps. aeruginosa* as test organism, Kataoka et al. [104] found a minor decrease in cellular yield (10 to 20%) caused by oscillations in dissolved oxygen (between approx. 10 and 60 ppm O_2) at rather short periods (about 1 min) only. For longer cycles cellular yields were the same as for steady state operation at low dissolved oxygen concentrations (7 to 10 ppm O_2); on the other hand, both cellular yield and specific growth were decreased by a factor of two to four, when the process was operated at constantly high dissolved oxygen concentrations of 60 ppm. In experiments simulating high tower processes for effluent treatment, Takamatsu et al. [105] obtained slightly decreased cellular yields and distinctly higher growth rates by applying cyclic changes of total pressure between one and 3 to 8 bar, though with dissolved oxygen far from saturation. Neither they nor Kataoka et al. [104] searched for metabolic oscillations.

On effects of cycling oxygen tensions on microbial synthesis of antibiotics, only a study by Vardar and Lilly [103] has been published. These authors subjected penicillin producing cultures of *Penicillium chrysogenum* to sinusoidal changes of total pressure with periods of 2 min. In this way they generated cyclic changes in dissolved oxygen concentration between 37% and 23% air saturation of the culture medium. These oscillations led to an approx. 30% decrease in specific penicillin production rate compared to penicillin production at a constant dissolved oxygen concentration of 30%

Table 6. Penicillin production at constant and cycling dissolved oxygen concentration DO, (Vardar and Lilly [103]). Specific penicillin production rate by *Penicillium chrysogenum* ($U\ g^{-1}\ h^{-1}$)[a]

at constant DO[b] (30% saturation)		at cycling DO[b] (between 37% and 23% saturation)
7400	→	5300
		↓
		5250
7450	→	
		5340
7500	←	
	→	5150
7350	←	

[a] penicillin units per g cell mass and h;
[b] dissolved oxygen concentration, saturation related to 1 bar air

air saturation (= mean oxygen concentration during cycling). This effect was reversible (cf. Table 6). The results demonstrate the unfavourable influence of fluctuating environmental conditions on process performance in large bioreactors.

6 Conclusions

From the influence of total and partial pressures on aerobic microbial cultures, the effects caused by oxygen partial pressure are by far the most important, whereas those of carbon dioxide partial pressure are of minor relevance. Compared to these partial pressures the effect of mere total pressure can be practically neglected. Essential parameters are not partial pressures, but actual concentrations of oxygen and carbon dioxide in the liquid medium. As to oxygen, the general tendency is evident, that higher concentrations from about 1 bar O_2 saturation onwards will affect microbial cultures negatively, leading to inhibition of growth, decrease in product formation and lower cell yield. In a few cases, however, increased dissolved oxygen concentrations may improve process efficiency, i.e. in the formation of oxygenated products, such as carboxylic acids and hydroxy compounds, or in the production of single cell protein.

Considering the great number of possible causes for these effects, it is obvious, that presumable effects of oxygen on a specific process cannot be predicted. In order to avoid negative effects and to open possibilities for utilizing experimental work is necessary to detect such influences. This is particularly important in process development. A special aspect to be considered in this connection is the fact that environmental conditions in industrial bioreactors are quite different from those on laboratory scale [116,117]. This is especially true for dissolved oxygen concentration with regard both to its level and its local variations. Higher hydrostatic and total pressures in industrial bioreactors increase gas solubility; pressure gradients and limitations in mixing and mass transfer in large bioreactors produce local differences in dissolved oxygen concentrations [118] (cf. Sects. 2.2 and 5.3). Possible effects of these changes in environmental conditions on performance of microbial processes should be investigated at an early stage of process development in order to avoid surprises at the start-up of the industrial unit. Information from experiments on static and dynamic effects of oxygen on the microbial culture is of course important for selecting the reactor type and for reactor scale-up. Also for improving existing processes, such investigations will be helpful, e.g. for the purpose of optimizing operating conditions.

The necessity of experimental studies directed to these problems has been emphasized by several authors [103,107,116,117]. However, the amount of such work is yet rather limited, possibly because specially suited laboratory outfit is required. This should be an opportunity for manufacturers to develop appropriate equipment.

7 Acknowledgements

The preparation of this manuscript would not have been possible without the grant of an "Akademie-Stipendium" by Stiftung Volkswagenwerk to one of us (U. Onken); this support is gratefully acknowledged. We should also like to thank present and former members of our team who helped us in various ways in completing the manuscript; in this respect we should like to mention especially Dr. Th. Jostmann (now Hüls AG), Dr. S. Kiese (now Wacker Chemie AG), Dipl.-Ing. D. Kaiser, Mrs. E. Rüther, Mrs. M. Verfürth, and B. Leeners.

8 Symbols

Symbol	Dimension	
c	kmol m^{-3}, g l^{-1}	concentration
D	h^{-1}	dilution rate
DOT	mbar	dissolved oxygen tension
f	bar	fugacity
I	kmol m^{-3}	ionic strength
K		equilibrium constant
k		rate constant
N	ml^{-1}	number of cells per volume
P	bar	total pressure
p	mbar	partial pressure
r	10^{-3} kmol m^{-3} h^{-1}	rate of reaction
RQ	—	respiration coefficient
T	°C	temperature
X	g l^{-1}	cell mass concentration
y	—	molar fraction in gas phase
Y	—	yield coefficient
μ	h^{-1}	specific growth rate
ν	h^{-1}	reproduction rate

Subscripts

i	component i
max	maximum
P	product
S	substrate
X	cell mass

Superscripts

G	gas phase
L	liquid phase

9 References

1. Gow JS, Littlehailes JD, Smith SRL, Walter FB (1975) Single cell protein from methanol: the development of the ICI process. In: Tannenbaum SR, Wang DIC (eds) 2nd Intern. Conf. on Single Cell Protein, MIT Press, Cambridge MA, p 370
2. Andrew SPS (1981) In: Proceedings of Inst. Mech. Eng. Conf. Future developments on process plant technology, 17
3. Westlake R (1986) Chem.-Ing.-Tech., 58: 934

4. Müller HG (1985) BIOHOCH Reactor of Hoechst AG. In: Rehm H-J, Reed G (eds) Biotechnology; Brauer H (ed) vol 2, VCH, Weinheim, p 561
5. Zlokarnik M (1985) Bayer Tower Biology, In: Rehm H-J, Reed G (eds) Biotechnology; Brauer H (ed) vol 2, VCH, Weinheim, p 556
6. Hines DA, Bailey M, Ousby JC, Roesler FC (1975) I. Chem. E. Symposium Series No. 41, D1
7. ZoBell CE, Johnson FH (1949) J. Bacteriol. 57: 179
8. ZoBell CE, Cobet AB (1962) J. Bacteriology 84: 1228
9. Pope DH, Berger LR (1973) Arch. Mikrobiol. 93: 367
10. Marquis RE, Keller DM (1975) J. Bacteriology 122: 575
11. Albright LJ (1975) Can. J. Microbiol. 21: 1406
12. Yayanos AA, van Boxtel R, Dietz AS (1983) Appl. Environ. Microbiol. 46: 1357
13. Thom SR, Marquis RE (1984) Appl. Environ. Microbiol. 47: 780
14. ZoBell CE, Oppenheimer CH (1950) J. Bacteriol. 60: 771
15. Rutberg L (1964) Acta Pathol. Microbiol. Scand. 61: 81
16. ZoBell CE (1970) In: Zimmermann AM (ed) High pressure effects on cellular processes, Academic, New York, p 85
17. Kinne O (1972) In: Kinne O (ed) Marine ecology, Wiley Interscience, London, p 1323
18. Marquis RE (1976) Adv. Microb. Physiol. 14: 159
19. Sato S, Mukataka S, Kataoka H, Takahashi J (1984) J. Ferment. Technol. 62: 71
20. Onken U, Jostmann T, Weiland P (1984) 3rd Europ. Congr. Biotechnol. Abstr. papers III, VCH, Weinheim, p 481
21. Onken U, Jostmann T (1984) Biotechnol. Lett. 6: 413
22. Hedén CG, Malmborg AS (1961) Sci. Repts. Ist. Super. Sanitá 1: 213
23. Othmer DF (1976) Process Biochem. 11: 8
24. Kämpf HJ (1977) Chem. Ind. 29: 329
25. Schumpe A, Quicker G, Deckwer WD (1982) Adv. Biochem. Eng. 24: 1
26. Schumpe A (1985) Gas solubilities in biomedia. In: Rehm HJ, Reed G (eds) Biotechnology, Brauer H (ed) vol 2, VCH, Weinheim, p 159
27. Freier RK (1978) Aqueous solutions, vol 3, de Gruyter, Berlin, p 17
28. King AD, Nagel CW (1967) J. Food Sci. 32: 575
29. Bylinkina ES, Nikitina TS, Biryukov VV, Cherkasova ON (1973) Biotechnol. Bioeng. Symp. 4: 197
30. Ishizaki A, Shibai H, Hirose Y, Shiro T (1973) Agr. Biol. Chem. 37: 99, 107
31. Ghandi AP, Kjaergaard L (1975) Biotechnol. Bioeng. 17: 1109
32. Chen SL, Gutmanis F (1976) Biotechnol. Bioeng. 18: 1455
33. Gill CO, Tan KH (1979) Appl. Environ. Microbiol. 38: 237
34. Puhar E, Lorencez I, Fiechter A (1983) Eur. J. Appl. Microbiol. Biotechnol. 18; 131
35. Jostmann T (1986) Einfluß des Gesamtdrucks und des Sauerstoffpartialdrucks auf das Wachstum des Bakteriums *Ps. fluorescens* im Druck-Schlaufenreaktor, Thesis, Dortmund University
36. Ho CS, Smith MD (1986) Biotechnol. Bioeng. 28; 668
37. Pirt SJ, Mancini B (1975) J. Appl. Chem. Biotechnol. 25: 781
38. Zajic JE, Liu FS (1969) Dev. Ind. Microbiol. 11: 350
39. Sturm FI, Hurwitz SA, Deming JW, Kelly RM (1987) Biotechnol. Bioeng. 29: 1066
40. Molin G (1987) Proc. 4th Europ. Congr. Biotechnol. 3; 49
41. King AD, Nagel CW (1975) J. Food Sci. 40: 362
42. Kobayashi K, Ikeda S, Hishinuma K, Hirose Y, Okada H (1972) Agr. Biol. Chem. 36: 961
43. Kretschmer A, Wagner F (1980) Eur. J. Appl. Microbiol. Biotechnol. 10: 41
44. May OE, Herrick HT, Moyer AI, Wells PA (1934) Ind. Eng. Chem. 26: 575
45. Wells PA, Moyer AI, Stubbs JJ, Herrick HT (1937) Ind. Eng. Chem. 29: 653
46. Träger M, Müller U, Onken U (1987) Chem.-Ing.-Tech. 59: 939
47. Oosterhuis NMG, Groesbeek NM, Kossen NWF, Schenk ES (1985) Appl. Microbiol. Biotechnol. 21: 42
48. Gottlieb SF (1966) J. Bacteriol. 92: 1021
49. Brown OR, Silverberg RG, Huggett DO (1968) Appl. Microbiol. 16: 260
50. Brown OR, Huggett DO (1968) Appl. Microbiol. 16: 476
51. Gottlieb SF, Pakman LM (1968) J. Bacteriol. 95: 1003
52. Gottlieb SF (1971) Ann. Rev. Microbiol. 25: 111

53. MacLennan DG, Ousby JC, Vasey RB, Cotton NT (1971) J. Gen. Microbiol. 69: 395
54. Matsumura M, Kobayashi J (1980) J. Ferment. Technol. 58: 552
55. Brunker RL, Brown OR (1971) Microbios 4: 193
56. Onken U, Kiese S, Jostmann T (1984) Biotechnol. Lett. 6: 283
57. Onken U, Buchholz R, Riethus M (to be published)
58. Liefke E, Onken U (to be published)
59. Harrison DEF, MacLennan DG, Pirt SJ (1969) In: Perlman D (ed) Recent advances in fermentation technology. Academic, New York, p 117
60. Hartmeier W, Bronn WK, Dellweg H (1971) Chem.-Ing.-Tech. 43: 76
61. Moore B, Williams RS (1909) Biochem. J. 4: 177
62. Moore B, Williams RS (1910) Biochem. J. 5: 181
63. Adams A, (1912) Biochem. J. 7: 297
64. Akashi K, Shibai H, Hirose Y (1979) J. Ferment. Technol. 57: 317
65. Páca J, Gregr V (1979) Biotechnol. Bioeng. 21: 1827
66. Liefke E (1988) Kultivierung aerober Bakterien bei erhöhtem Sauerstoffpartialdruck als verfahrenstechnische Möglichkeit zur Beeinflussung von Wachstum und Produktbildung, Thesis, Dortmund University
67. Webley DM (1954) J. Gen. Microbiol. 11: 114
68. Lee F-JS, Hassan HM (1987) Appl. Microbiol. Biotechnol. 26: 531
69. Clark DS, Lentz CP (1961) Can. J. Microbiol. 7: 447
70. Khan AH, Ghose TK (1973) J. Ferment. Technol. 51: 734
71. Kristiansen B, Sinclair CG (1979) Biotechnol. Bioeng. 21: 297
72. Okoshi H, Sato S, Mukataka S, Takahashi J (1987) Agr. Biol. Chem. 51: 257
73. Shibai H, Ishizaki A, Mizuno H, Hirose Y (1972) Agr. Biol. Chem. 37: 91
74. Yamada S, Mitsuru W, Chibata I (1978) J. Ferment. Technol. 56: 29
75. Buckland BC, Lilly MD, Dunnil P (1976) Biotechnol. Bioeng. 18: 601
76. Halliwell B (1977) Dechema Monogr. 81: 1
77. Halliwell B (1982) Trends Biochem. Sci. 7: 270
78. Fridovich I (1978) Science 201: 875
79. Brown OR, Yein F, Boehme D, Foudin L., Song CS (1979) Biochem. Biophys. Res. Comm. 91: 982
80. Robb SM (1966) J. Gen. Microbiol. 45: 17
81. Caldwell J (1965) Nature (London) 206: 321
82. Gifford GD, Pritchard GG (1969) J. Gen. Microbiol. 56; 143
83. Sies H (1986) Angew. Chem. Int. Ed. Engl. 25: 1058
84. Sies H (1984) Detoxification of oxygen free radicals. In: Bors W, Saran M, Tait D (eds) Oxygen radicals in chemistry and biology, de Gruyter, Berlin, p 653
85. Jones DP (1985) The role of oxygen concentration in oxidative stress. In: Sies H (ed) Oxidative stress, Academic, London, p 151
86. Brown OR, Yein F (1978) Biochem. Biophys. Res. Comm. 85: 1219
87. Boehme DE, Vincent K, Brown OR (1976) Nature 262: 418
88. Haugaard N (1968) Physiol. Rev. 48: 311
89. Barron ESG (1955) Arch. Biochem. Biophys. 59: 502
90. Williams RJ (1981) Oxygen and Life: An introduction, in: Oxygen and Life, Congresses I. Royal Soc. Chem., Birmingham 1980, Whitstable Litho, p 18
91. Haber F, Weiss J (1934) Proc. Roy. Soc. Acad. 147: 332
92. Cohen G (1985) The Fenton reaction. In: Greenwald RA (ed) Handbook of methods for oxygen radical research, CRC Press, Boca Raton
93. de Groot H, Noll T (1987) Chemistry and Physics of Lipids 44; 209
94. Halliwell B, Gutteridge JMC (1986) Arch. Biochem. Biophys. 246: 501
95. Dahl TA, Midden WR, Hartman PE (1987) Photochem. Photobiol. 46: 345
96. Wolff SP, Garner A, Dean RT (1986) Trends Biochem. Sci. 11: 27
97. Bruynickx WJ, Mason HS, Morse SA (1978) Nature 274: 606
98. Farr SB, D'Avi R, Touati D (1986) Proc. Natl. Acad. Sci. USA 83: 8268
99. Gebicki JM, Bielski BHJ (1981) J. Am. Chem. Soc. 103: 2020
100. Dalton H, Postgat JR (1969) J. Gen. Microbiol. 54: 463

101. Lowe DJ, Thornerley RNF, Smith BE (1985) Nitrogenase. In: Harrison PM (ed) Metalloproteins I, VCH, Weinheim, p 207
102. Dingler C, Oelze J (1985) Arch. Microbiol. 141: 80
103. Vardar F, Lilly MD (1982) Eur. J. Appl. Microbiol. Biotechnol. 14: 203
104. Kataoka H, Sato S, Mukataka S, Namiki A, Yoshimura K, Takahashi J (1986) Biotechnol. Bioeng. 28: 663
105. Takamatsu T, Shioya S, Nakatani H, Fujimoto T, Kawasugi T (1981) Adv. Biotechnol. 1: Pergamon, Oxford, p 581
106. Soni VK, Ghose TK (1974) J. Ferment. Technol. 52: 551
107. Katinger HD (1976) Eur. J. Appl. Microbiol. 3; 103
108. Gifford GD, Pritchard GG (1969) J. Gen. Microbiol. 56: 143
109. Gregory BM, Fridovich I (1973) J. Bacteriol. 114: 1193
110. Páca J (1980) Eur. J. Appl. Microbiol. Biotechnol. 9: 93
111. Harrison DEF, Pirt SJ (1967) J. Gen. Microbiol. 46: 193
112. Harrison DEF (1976) Adv. Microb. Physiol. 19: 243
113. Harrison DEF, Topiwala HH (1974) Adv. Biochem. Eng. 3: 167
114. Sokolov DP, Lirova SA, Sokolova EA (1983) Microbiology (USSR) 52: 715
115. Lirova SA, Sokolov DP, Senyushkin AA, Berestennikova ND, Rabotnova IL (1986) Microbiology (USSR) 55: 104
116. Sweere APJ, Luyben KChAM, Kossen NWF (1987) Enzyme Microb. Technol. 9: 386
117. Vardar F (1983) Process Biochem. 18: 21
118. Onken U, Weiland P (1983) Adv. Biotechnol. Proc. 1: 67
119. Danckwerts PV (1970) Gas-liquid reactions, McGraw-Hill, New York, p 239
120. Kiese S (1982) Untersuchung des Druckeinflusses auf das aerobe Wachstum einer Hefe im Airliftfermenter, Thesis, Dortmund University